LIVING FORESTS

HEINRICH GOHL   LIVING

TEXT Dr E. KREBS

TRANSLATED BY FREDERICK AND CHRISTINE CROWLEY

FORESTS

KAYE AND WARD   OXFORD UNIVERSITY PRESS

First published in Great Britain by
Kaye & Ward Ltd
21 New Street, London EC2M 4NT
1975

First published in the USA by
Oxford University Press Inc.
200 Madison Avenue
New York, NY 10016
1975

ISBN 0 7182 1100 6 (Great Britain)
ISBN 019 519802-6 (USA)
Library of Congress Catalog Card Number 74-20640 (USA)

CONTENTS

The forest is a thing of mystery, incomprehensible by its very nature. It can only be experienced — felt. The language of the forest has to be learnt; it is a language of wonder, stillness and reverence for the intangible.

Millions of years before man appeared on Earth the forest already stood. It would today have been more extensive, more natural, more abundant and richer in flora and fauna had it been spared the interference of man. The forest has no need of man; it can exist without him. But man cannot live without the forest. Where the forest is destroyed, there dies with it all life-giving Nature; man's environment becomes uninhabitable.

The forest is not only of the greatest importance to the countryside's ecology, it also has a most significant impact upon man's intellectual and spiritual life. Nature holds innumerable secrets that we have yet to unravel. The further we penetrate, the more riddles arise and the more remote the final secret becomes.

The Earth's forests are of inconceivable variety. They produce the wood we need in one form or another to live; they offer protection against many of Nature's dangers; they protect soil, water, air and countryside. Above all they make an overwhelming impression on the people fortunate enough to experience them. It is a comforting thought that the forest is still available to us. In areas developed by man the forest embodies all that is permanent, living and natural. Only in the forest do we feel Nature's diversity. The forest lives on, mysterious and wonderful. Where in today's world of technology can man come nearer to the primeval power of Nature than in the forest? The forest is eternally alive, continually recreating. In the Iliad, Homer compares human life and the human race to the forest: ''As the leaves in the forest, so is the human race; the wind scatters the leaves on the ground, others are put forth by the verdant forest when Spring burgeons once again. Likewise the human race: one man grows and another disappears.''

We shall attempt to describe the innermost nature of the forest. Man cannot be excluded from the discussion, since on the one hand he was himself responsible for the development of forest and countryside over vast areas, and on the other hand the forest determines the future of mankind. The man—forest dualism extends like a red thread through the history of all races. Man was born into this world with the task of living here, of utilizing the countryside, of shaping it, but also with the task of preserving it. Unfortunately he has abused the Earth, impaired its natural beauty, indeed even destroyed it in many places. Through technology and science we are able today to manufacture efficient machinery; scientists have invented atomic energy, numerous synthetic materials, new medicaments and highly effective poisons. However, we do not have the ability to create a living flower, a small sentient animal or a tiny seed in which the rich genes of the mother tree lie dormant with those of the unknown father tree from which the wind blew the male pollen. May it always remain so! Let man realize at long last that he has no power to rise above Nature. He must come to see that, in spite of science and technology, he knows very little, and that the tiniest work of Nature is more magnificent and more mysterious than the most powerful machine he can build. The final and deepest secrets of Creation will always remain hidden from us. For this reason, man should always be aware of the threshold that for ethical reasons he is not entitled to cross.

Love of the forests begins in childhood, when the infant, with his parents, visits the mysterious forest and is made aware of its beauties and secrets. This love deepens in the growing man and he absorbs the magnitude and the beauty with increasing amazement and reverence. "This has made me love and trust the forest: it threads its way through all my childhood dreams back to my earliest memories; it is the friend of my youth and has remained my secret refuge into the autumn of life" (Augustin Wibbelt).

How the forest came into being

The world has a history going back four or five milliard years. Revolutionary changes occurred, primeval forces had their consequences; violent upheavals followed one another until the end of the Ice Age. Much is still unknown; nevertheless, research has succeeded in illuminating some part of the picture of the creation of our Earth in the most distant past. The remains of flora and fauna that lived in those ages were embedded by sedimentation in primeval bogs and can be related to the different geological formations. Accurate age-determination has recently become possible with the aid of modern atomic physics.

Trees and forests were also part of these early stages. Present-day flora and the first tree species can be traced back to tertiary primeval flora and post-tertiary development. On land, the forest was always the strongest and most widespread vegetation group. "The first conqueror of the world was the forest" (A. Metternich, 1947).

Although post-Ice Age development had a special effect upon the present composition of the forest, earlier conditions were not without importance. The geological periods given below are based on the account of A. Holares, 1960.

The Earth's antiquity (Palaeozoic, 600–225 million years ago).

A picture of the first tree-like plants may be seen in the fossils of an epoch in the world's history more than 400 million years ago (Devonian). There existed, initially, low forms of vasculiferous plants with no needles or leaves, which were propagated by spores. About 300 million years ago, during the Carboniferous period, there evolved great swamp forests with strange, tree-like lycopods, ceterachs, lepidodendrons and giant horse-tails up to sixty feet high which formed gigantic seams of coal. Arizona has vast areas of petrified forest. Fifty million years later (Permian), older species of flora and fauna became extinct,

18

and conifers increasingly expanded. The first gingko trees appeared, a primitive form of conifer still extant today. Conifers and tree-ferns formed the main component of Stone Age fossil flora until the newer layers of the Triassic formations.

The Earth's middle age (Mesozoic, 225–70 million years ago).

During the first part of this period (Triassic), a flora existed similar to that of a sub-tropical region. Horse-tail, fern and conifer predominated, along with palm-like trees such as are still found today in the tropical forests of Africa and America. During the middle of this period (Jurassic), when the continents sank and more areas became flooded, the flora of a tropical region predominated. The first gymnospermous plants appeared. At the end of the Jurassic period, the gigantic continental block began to split into separate continents. Only this can explain the propagation of certain plant families. At the end of this period (Cretaceous), during the great expansion of the oceans, the beginning of the plication of the Alps and the rising of the Rocky Mountains and Andes, the greatest leap in plant development took place. As well as ferns, horse-tail and conifer, deciduous trees, shrubs and plants developed in great abundance.

The Earth's present (Neozoic – began 70 million years ago)

This period was of special importance for present day conditions.
Tertiary period (70–1 million years ago). Significant upheavals took place in the Earth's history until the beginning of the Ice Age. The Alps, the Pyrenees, the Carpathians and the Central African mountain region were folded upwards to their maximum height. The North American continent broke away from

Europe. In East Africa, the famous fault-block depressions occurred. Tertiary lignite contains fossils of a rich tropical and sub-tropical jungle vegetation. Apart from mammoth trees, swamp cypresses and eucalyptus, there grew breadfruit, camphor, cinnamon, fig, tulip trees and laurel, and there were many of today's deciduous species, such as chestnut, oak and palm. Molasse deposits have traces of more than 1500 plant species, including — in the European region as far as Spitzbergen and Northern Greenland — numerous tree species that are still growing in North America and Japan. At the end of the Tertiary period, the entire northern hemisphere had a more or less uniform climate and was covered with a similar flora, rich in species. Almost 90% of our present-day flora and fauna existed then.

Quaternary period. During the last, very short chapter of the Earth's history, about one million years ago, the Ice Age catastrophe befell the Earth's flora and fauna. Over the whole northern hemisphere a continuous covering of ice pushed its way far southwards. Great ice masses flowed from the mountain massifs and combined to form a wide belt of ice. Vegetation had to yield to this invasion by the ice masses. It retreated into refuges that remained ice-free. In Europe a mighty barrier to the retreat of the flora was formed by the Pyrenees, the Alps and the Carpathians. In particular, those tree species that needed warmth were unable to overcome this barrier. Sub-tropical flora, therefore, disappeared from Europe forever.

Between the ice front of the northern glacial formations and the edge of the Alpine glaciers there remained a plain about 250 miles wide, ice-free and with a sparse arctic-alpine flora of dwarf birch and dwarf willow. Classical theory posits four great glaciation phases, with warmer intermediate periods. Each glacial retreat in the intermediate periods was followed by an advance of vegetation; but, with renewed expansion of the glaciers, vegetation retreated into its refuges. The end of the Ice Age in Europe occurred from about 10,000—8,000 BC. The retreating glaciers left a barren wasteland. However, in less exposed areas life had been preserved, and from here the green cloak of vegetation gradually took possession of the bare no-man's-land insofar as climatic frontiers did not of themselves prevent re-conquest. The European mountain ranges, in the same way as they had made more difficult the retreat of the flora and fauna from the advancing ice flows, also prevented or delayed their reappearance in the post-glacial period.

It is fascinating to follow this reappearance in space and time, and the routes it took during the post-glacial period. Nature has itself kept a diary of this, from which the expert can deduce in broad terms the post-glacial development of climate and vegetation and the process of afforestation. The most important means of deducing this is by pollen analysis. The small male pollen grains of the forest trees have been preserved over thousands of years in the peat layers of moors and bogs and in the glacial clay. From the pollen ratio of successive peat layers, the composition of the forest tree species on the moors can be deduced. In the interpretation of pollen diagrams, however, various sources of error must be considered. For example, there are several forest trees and shrubs whose pollens are not easily preserved, and consequently these species are usually missing from pollen diagrams, although their existence has been established by findings of fossil wood. On the other hand, frequently-flowering trees with high production of pollen are often over-represented in the diagrams. In spite of this, pollen spectra present in their general structure a fairly clear picture, and make it possible for us to illuminate prehistory that would otherwise remain completely obscured from a geo-botanical viewpoint. The decay of radio-activity in the carbon of organic residues has made it possible in the last twenty years to determine fairly accurately the date of individual layers.

The occupation of the land as the ice retreated gives the impression of a strategic campaign. The glaciers, abandoning the field, were first followed by spearhead troops: low-growing, unassuming pioneer plants such as lichen, juniper and dwarf birch like those found today in the far North and in the high Alps. These first colonizers pushed their roots through the stony soil, burst it open, extracted the sparse nutrients and formed the first traces of humus. They prepared the ground for subsequent, more demanding immigrants, who settled under the shelter of the pioneers and pushed them out. Forest trees followed from various directions, their way always prepared by pioneers. They came mainly from the glacial refuges of the Iberian peninsula, Italy and the Balkans. Their progress rested upon their ability to produce light seeds in abundance, which were sent ahead by the wind.

Taking into consideration the dislocations in time, it is possible to arrive at the following broad development in Central Europe:

Late Ice Age (late glacial period, 10,000–8,300 BC). Last retreat of the glaciers 12,000 years ago; renewed brief advance about 11,000 years ago. Birch and common fir forests predominate; forest landscape similar to that of present-day Lapland.

Pre-tropical period (Preboreal, 8,300–6,500 years BC). End of the last Ice Age, about 8,000 years BC. Climate warmer and drier than today. The late-glacial birch and common fir vegetation remains for a further 1,000 years. The hazel tree then begins to expand and reaches the height of its development about 7,500 years BC.

Early tropical period (Boreal, 6,500–5,500 BC). With the temperatures higher than today, mixed oak forests expand with warmth-loving companions such as elm, lime and maple, with a rich shrub layer. Towards the end of this period they are already extending upwards into the present-day high-mountain level in the Alpine range.

Old tropical period (Old Atlanticum, 5,500—4,000 BC). Oceanic, humid climate. The mixed oak forest has reached the peak of its expansion. Beech and spruce advance into existing forests.

New tropical period (New Atlanticum, 4,000—2,500 BC). Strongest expansion of beech tree in Central Europe. In the higher altitudes of the Northern Alps, the red fir is beginning to form the mighty sub-Alpine belt of pine forests.

Late tropical period (Sub-boreal, 2,500—800 BC). Around 1,000 BC the climate is becoming cooler and precipitation is increasing. Mixed oak forests remain in the lowlands, while beech and spruce increasingly penetrate into the mountain valleys.

Post-tropical period (Sub-Atlanticum, 800 BC until today). Formation of present-day forest communities. Man begins to influence forest conditions.

Development on the North American continent took a substantially different course. During the advance of the ice masses, the mountain chains running North-South made it possible for the flora and fauna to retreat almost unhindered into warmer zones. Similarly, there was no restraint upon their return during the post-glacial period. The flora is consequently not impoverished to the same extent as in Europe. Whereas only about fifty species of tree exist in Cisalpine Europe, there are 150 in Canada and 850 in the United States.

Eastern Asia to a large extent escaped glaciation, and therefore today possesses the forests that are most rich in species. This explains why species of the Tertiary period, such as mangrove, palm, fig, mimosa, gingko, magnolia, etc., are found today only in Southern Europe, apart from their main habitat in Japan, China and North America.

The world map of forest zones shows that two large belts of forest encompass the globe. One is the equatorial belt of tropical rain forest, jungle and arid forest. More than 10,000 different species of tree are known in the tropical rain forest. The arid forests of the tropics, on the other hand, have few species, since only specially adapted species can withstand the long dry periods. The other forest belt is located in the northern temperate zone and includes the gigantic boreal coniferous forest belt that extends as far as the 69th parallel; in the South it meets the forests of the temperate and sub-tropical zones. The southern hemisphere contains nothing corresponding to the boreal belt of coniferous forest.

The variety of the species

The number of tree species upon the Earth is incredibly large. Conifers, however, are represented by only about 500 species. The multiplicity is much greater among deciduous trees; the precise number of species is not known, but it is estimated that the Amazon region alone has more than 3,000 species. According to H. Knuchel, about 33,000 tree species are known, of which only a few hundred are as yet valued for their timber.

The differences between extreme examples are fabulously varied and almost unbelievable. Whereas near the Polar timber line only bush-like trees can thrive, and in the arid regions only low woodland, there are eucalyptus trees in the Australian savannah reaching a height of 390 feet; one tree 508 feet high has in fact been reported. In the West of North America, Douglas firs of 426 feet are also known. The tallest coast redwood is said to be 442 feet high. The thickest trees in the Californian Sierra Nevada have a trunk diameter of up to 32 feet. These trees are of corresponding age. 3,500–4,000 years are recorded on the stumps of large redwoods. The thickest tree in the world is said to be a Mexican cypress in

Santa Maria de Tule, south of Oxaca; this has a trunk diameter of 39 feet and an estimated age of 2,000–5,000 years. In Africa, Adamson discovered the noble baobab, the monkey breadtree; he estimated that one of the giants he examined was 5,150 years old. In 1954, living bristle-cone pines were found in the White Mountains of California whose trunks displayed more than 4,000 rings, but which because of the unfavourable conditions (11,482 feet above sea level) were still only 32–39 feet high. Feininger came across a bristle-cone pine in Arizona whose age is estimated at 4,600 years. In the New Stone Age, more than 2,600 years before the birth of Christ, in Babylonian times, a grain of seed germinated there and grew into a tree which today is among the oldest living things on earth. When the pyramids were built in Egypt, this tree was already 1,400 years old. Another 1,300 years passed until, in the twenty-seventh century of its existence, the birth of Christ was proclaimed. The tree still bears foliage, though its trunk is gnarled and its bark is worn away by centuries of blizzards and sandstorms. But a strip of bark still lives on one side of the trunk, and year by year the sap rises from a few roots to the few branches that are still alive, adding at each growing period a wafer-thin annual ring to the wood on that side.

The various species of plants and trees with which Nature has equipped all countries do not, however, occur in unruly profusion. Each species has its special properties, its specific demands on soil, humidity, light and warmth. At any given place, the competitive power of the species that finds all its needs met there is strengthened, whereas the vitality of species less well adapted to the location is weakened. By an infinitely long process of elimination, plants with similar living requirements have gathered together into communities characterised by their sociological composition. We find, therefore, in some natural forest communities a multitude of tree and plant species, whose outstanding characteristic is the richness and variety of their mixture — a mixture influenced, of course, by their particular location. Fairly uniform societies, with relatively similar plant cover, are found in large forest areas with the same climatic and edaphic conditions. Natural forest communities may also be organized in discrete areas in a varied mosaic of ecologically dissimilar habitats. In extreme, unfavourable localities, in the boreal pine forest belt, in mountains and on poor soil, the forest communities are usually poor in species, since here only the specialist of modest requirements can survive. The forests of the warm Mediterranean countries and of tropical regions are, however, infinitely richer in species. Here natural advantages offer to many species an almost limitless supply of what they need for their existence.

It is fascinating to follow the development of vegetation, beginning with the first primitive communities on the raw soil of the post-glacial period, and proceeding, by continuous development, through a succession of intermediate communities to the climatically and edaphically conditioned present-day communities. Gradual change is still taking place in vegetation, caused by the development of climate and soil. However, this change is so long-term that the successive stages reached by vegetation must appear to be permanent communities to us short-lived humans.

When we follow this breath-taking development, showing how, over immeasurably long periods in a never-ending sequence, plants and animals appeared and became extinct, when we comprehend the fertility of Nature's invention by which it continually forges new forms, we sense how small and insignificant we are in relation to Nature's omnipotence. In the course of many hundreds of millions of years, Nature, according to its own laws, created life, from the simplest forms to the highest levels. "In its tentative attempts, life seems to have tried everything; it is all the more strange that it has always tried in the same direction" (Teilhard de Chardin).

34

Growth of the tree and the forest

Although many of the details of growth are known, Nature's greatest wonders are still hidden in this process. The wood substance of a tree consists of 99.6% organic matter which the tree has itself built up by assimilation. Only a bare 0.4% is inorganic substance originating from the soil. Through the stomata of leaves and needles, trees absorb carbon dioxide from the air and, with water from the soil, form sugar in the cell plasma, thus releasing excess oxygen. The sugar is used to form cells or is converted into starch and stored in the trunk. Through the breathing of the plant — the reverse process — about one-fifth of the sugar produced is utilized again.

The assimilation of green plants can be expressed by a simple formula:

$$6\,CO_2 + 6\,H_2O \xrightarrow{\text{Light}} C_6H_{12}O_6 + 6\,O_2$$

From this it can be calculated that green plants, with the help of sunlight, produce, from 264g of carbon dioxide and 108g of water, 180g of glucose; in the process they expend 673 calories of energy and release 192g of oxygen. We do not know why Nature has arranged things in this way or why the structural composition of the organic substances is so complicated.

Only about 1% of solar radiation energy is used in the photosynthesis of the green plant cover. Conifers are generally photosynthetically more effective than deciduous trees, although outside the growing period the latter are also able to assimilate above the temperature threshold of 4°C. Assimilation takes place only during daylight and only in green plants with chlorophyll (leaf-green), which in the astonishing process of photosynthesis acts as a catalyst and is continuously formed in the needles and leaves. Colourless plants are unable to assimilate solar energy, but must be content with a parasitic life on other plants or catabolic substances. Photosynthesis in green plants is the only method of producing matter in which natural solar energy is used and the air is at the same time renewed. This is particularly impressive in those trees where the process has continued in the same tree for decades, centuries and millennia, even when the inner trunk has long been dead wood.

Fertile earth

The forest floor is not merely a place for the trees to stand on; it is also the soil from which they extract water and nutrient salts. It is, therefore, understandable that the growth of trees and forests is more intensive on humid soil rich in nutrients than on poor, dry soil, and that assimilation intensity is much greater in a warm, humid climate than in cold or arid regions. The basis of plant life is water. The soil is a mighty accumulator of water and mineral nutrient salts. Day and night during the growing period plants draw from the soil water that is replaced by precipitation. Through the pipelines — vessels in deciduous trees, tracheae in conifers — that run inside roots, trunk and branches, water containing the dissolved mineral substances is led to the assimilating leaves and needles. A relatively small amount of water is chemically combined in the production of matter; the greater part is returned to the atmosphere by the evaporating crown. Forests are large water consumers during the growing period. An old beech wood can evaporate up to 2,600 gallons of water per acre a day during the summer, and a forest of poplars up to 4,300 gallons. In the case of the large trees, the water must rise to a height of more than 330 feet. By evaporation, suction and lifting power is created, which is conveyed right through the trunk down into the roots. The cohesive force that lifts water from the soil to the very tip of large trees amounts to as much as 30 atmospheres.

The root network of a tree is an enormous system whose ramifications may perhaps be compared with the tree's crown. Depending on the type of soil, an average-size common fir will drive into the soil one or more main roots of approximately 10–26 feet in length. From these extend hundreds of secondary roots. The branching becomes ever wider and finer; the final fibrous roots are hair-thin and only a few tenths of an inch in length. The total length of such a root system can amount to thousands of miles. Cell division takes place continuously at the tip of each root, while the young cells in the rear expand by stretching until they are ten times their original length within a few hours. This is why the fine roots, protected by the root cap, penetrate deeper and deeper into the soil, continuing to drill in their search for water and nutrients.

In the nourishment of most forest trees certain species of fungi which live together with the tree roots in a life partnership, referred to as mycorrhiza, play a crucial part. It has been demonstrated that the absorption of matter by these trees is not effected directly but in a round-about way via the mycorrhizae, which draw the mineral salts from the soil with their hyphae. In conifers particularly, but also in many other deciduous trees, the fungus often completely encloses the young root tips (ectotrophic mycorrhiza) that are responsible for absorbing the nutrients. In herbaceous plants and also in many lignified plants the fungi penetrate into the root web (endotrophic mycorrhiza). Many researchers have come to the conclusion that plants with fungi-free roots are in practice non-existent. In general, young plants germinating in the forest soil develop a fungoid cloak around the root tips even during the first year. The mycorrhizae react in a strangely sensitive way to unfavourable soil conditions, in particular to unsuitable manuring and to chemical toxins. Very little research has as yet been done into the many aspects of this strange symbiotic relationship between plants and root fungi. On the one hand, the fungus supplies the plant with valuable nutrient substances, but on the other hand the plant can become weakened by a fungoid growth that is too extensive.

Annual rings tell a story

The thickness of growth of trunk, branches and roots is effected within the formative tissue (cambium) located between wood and bark. In this tissue layer, wood cells are formed inwards and bark cells outwards by continuous cell division during the growing period. Division takes place with great rapidity. "If someone were to count cells continuously night and day at about 200 per minute he would reach 105 million a year without keeping pace with the cell growth of a tree" (A. Bernatzki).

Wood cells vary widely in their form according to their function (longitudinally directed tracheae and parenchyma cells in conifers; vessels, tracheae and parenchyma cells in deciduous trees, as well as radial silver grain tracheae and parenchyma cells, that is two cell systems overlapping one another, in the trunk). In the spring the cambium forms loosely constructed cells with soft, light wood and many conducting elements for the tempestuous onset of the sap flow. In late summer on the other hand, thick-walled cells are formed with hard, dark wood, which mainly provide strength. In old trees, the outer alburnum zone usually still conducts water, while the inner section of the wood body gradually dies off and is no longer of any physiological use. In the cambium, however, cell division progresses, and one of the greatest marvels is that the formative tissue of the oldest trees known is still able to live after more than 4,000 years and continues to grow, although the trunk interior has long been no more than dead, supporting wood tissue. For this reason trees can live much longer than man and the animals, in whom living and dead tissue cannot exist simultaneously.

The light, wide-meshed spring wood and the outer, enclosing, dark, hard autumn wood together make an annual ring, since this is formed during each growing period in areas with a seasonal change. On good soil wide annual rings

39

develop, and on bad soil narrower rings. Trunks in mountainous regions and in northern areas where the growing periods are short have very close rings, while those in warm, rainy areas are broad-ringed. However, even in the same tree the rings are usually of uneven width. Repressive periods can be recognised by the narrow rings, when the tree had to be content with little light in the shadow of its competitors; later, broader rings formed once again when it was freed from its harassing neighbours. In sunny, humid years, the annual rings become broader; when drought prevails, growth is slight. Strong attack by pests and fungoid diseases or excessive seeding that weakened the tree are characterised by narrow rings. Thus the tree tells the story of its life, its good and bad times. Only in tropical rain forests, where neither winter nor drought interrupts growth, are annual rings lacking in the wood.

Since there is a close connection between weather conditions and the width of the rings, it has been possible in Germany to establish from numerous trunk layers complete annual rings of beech and fir as far back as the early Middle Ages, as well as an oak chronology covering a thousand years. It is possible by this means to date ancient wooden buildings and wood-carvings, provided a section is available in which at least fifty annual rings are recognizable. At the University of Tucson, an annual ring chronology stretching back 7,000 years has been produced.

Prodigal Nature

Inscrutable Nature is prodigal in ensuring the continuing existence of the various species of tree. The differing solutions that Nature has found for the culture of flowers and the mechanism of their fertilization are indeed surprising. In many deciduous trees the pollination of the female flower is performed by

40

insects, which are attracted by scent, colour or protein-rich pollen. In other deciduous trees and in most conifers the male pollen is transferred to the female stigma by the wind. The tiny pollen grains have an amazing capacity to fly. 50,000–60,000 beech pollen grains weigh only one gramme. In a year when the seeds are particularly strong the anemophilae develop enormous quantities of pollen to meet the uncertainty of pollination. At Wassertümpeln, in the Zurich area, more than 20,000 grains of tree pollen per square yard of water surface were counted in 1950 (sulphur rain).

The lavishness of seed quantities also seems immeasurable. A fully-grown aspen produces some millions of seeds a year. In tree species which produce large seeds the quantity of seed is smaller, since a great deal of nutriment is needed for their formation. In good seed years, 50–90 cubic feet of additional growth per acre can be taken from a forest crop for seed formation. In spite of this, seed production is wasteful in the case of heavy-seed tree species. We have counted in beech woods during one year of growth up to 300 seedlings per square yard; that is, over 1 million per acre. In oak woods, a count resulted in almost 300,000 seedlings under just one large mother-tree.

The shape, size and weight of the seeds formed are also of infinite variety. 10,000 poplar seeds weigh only ·04 of an ounce, 1,000 acorns on the other hand weigh as much as 7–11 pounds. Thus, there are about 10 million seeds in each $2\frac{1}{4}$ pounds for the poplar and 200–300 seeds for the oak. There are species of conifer that form only short cones up to $\frac{1}{3}$ inch in length. Mammoth trees, the giants of more than 330 feet, develop relatively small cones only $2\frac{1}{2}$–3 inches long with 200–300 seed grains, whereas certain types of fir produce cones weighing up to $5\frac{1}{2}$ pounds ("widow-makers").

Nature also uses all possible means to spread the ripe seed. The heavy-seeded fruit falls directly to the ground and perhaps rolls a little way down a slope. It therefore grows only near the mother tree, provided it is not carried further by birds, mice, hamsters, squirrels or other animals. Other seeds, with a woolly flock or wings, are able to fly. Very light seeds can be transported hundreds of miles by the wind. Other seeds have sticky organs; they cling to the plumage or fur of an animal and allow themselves to be borne along until they are brushed off at random in one place or another. In mangrove forests in the flooded areas of tropical jungles, the seeds germinate on the mother tree so that the seedlings with their roots can perhaps gain a foothold elsewhere when carried along by the water.

Every year, or every few years, Nature thus broadcasts an abundance of seeds. By far the greater part of them is lost, and only a negligibly small number fall into a favourable bed for germination. Each grain of seed is provided with a small store of sugar, starch, fat and oil. Heavy seeds are given a large rucksack of provisions, but light seeds are only modestly equipped. These provisions must suffice to force out a little shoot and for a little root to reach nourishing soil. What happens when the provisions become exhausted before the fragile root finds water and nourishment, when a stone, a thick leaf or a small piece of wood prevents access to the soil? This is why the greater part of seedlings are destroyed — because of lack of nutriment, drought, frost and disease. Even the lucky ones are by no means safe; uncountable numbers are destroyed by weeds, lack of light or by being eaten by animals.

The harmonious community of the forest

Only a very few of the people who enter a forest consider what the forest in its whole complexity actually is. They usually allow themselves to be instinctively fascinated by the mysterious vastness of the space, the trees being the largest and most eye-catching members. But the forest is more than soil, trees and wood. Within the air space and root areas of the forest there lives an immense, indescribable abundance of flora and fauna, which together with the trees comprises the world of the forest.

The sum of all organisms in a living space (biotope) forms a community (biocoenosis). Living space and the living community together form a total structure (ecological system). The forests therefore are living communities rich in species, ecological systems difficult to survey as a whole. All animals and plants, even the most insignificant and tiny as well as the apparently harmful, have special assignments in the common destiny of the community, although these tasks may vary in their importance and may be known or as yet unknown. Each living thing is fitted in somehow and somewhere into the community's way of life, living with it and from it; everything is linked with everything else.

The pattern of such a living community is so interlaced that science has not yet succeeded in clarifying from every angle the affiliations and connections. It may well be impossible for us ever to extract all the secrets from this multifarious complex of relationships. No only do mutual ties exist between individual members in the living community of the forest, but a constant, merciless struggle also prevails. This is the struggle of the stronger against the weaker, of the larger against the smaller; and at the same time it is a struggle against unpredictable forces, against the laws of habitat and evolution, against the accident of advantage or of disadvantage from disease, a struggle according to the biological laws of selection. The struggle for life takes place not only visibly in the area of the crown, above the ground; it is fought with equal bitterness for water and nutriment, but this is hidden in the dark under the soil, in the root area. If we appreciate that under one square yard of forest floor there can be more than 500 yards of roots, and if we think of the innumerable mass of soil creatures who also take part in this struggle, we may then get some inkling of the even more entangled relationship of life in the soil.

The great house of the forest

Because of its great height, a forest is subdivided into different floors as in a house. A magnificent differentiation between the necessities of life of all members of the forest permits such a stratification in the vertical plane. In our forests, 2–5% of open-air light still prevails at the herbaceous layer. Whatever seeks to thrive at this level must, therefore, be able to tolerate shade. The enclosed space of the forest also has a special climate, differing considerably from the climate of the neighbouring open air.

Average temperature fluctuations in the forest are substantially lower during the day as well as over the year. Humidity is higher; the soil is always in shade; there is less wind. It is important, therefore, not to tear away the protective forest outskirts with their low-branched trees and enclosing shrub belt.

At ground-level in the interior of the forest moss, lichen, fungi, dried leaves and needles form a cover that protects the soil from drying out and becoming hard. Over this is an herbaceous layer made up of annual or perennial ferns, grasses and plants, which, depending upon the amount of light the crown allows to penetrate, is sometimes sparse and sometimes lush. Old tree stands with many gaps, sparse mountain forests and open woods in the North usually have a close undergrowth. The ground vegetation in young dense forests, in particular close pine forests that remain almost uniformly dark throughout the year, is on the other hand very poor indeed. The herbaceous layer, adapted in its pattern of species to the density of the stand, therefore reflects the light conditions in the region of the trunk base. The short-lived fungi population springs from the forest floor in all sorts of shapes and colours. Fungi are parasites. Most fungi spread through the forest floor with a web of fine fungus threads. Since they assist in decomposing dry branches, needles and dead wood, they are very important to the nutritive balance. For this reason even non-edible and poisonous fungi should not be wantonly destroyed.

One floor higher there is the shrub layer, well or poorly developed according to the light conditions. This layer is important in that it gives shade to the ground. At the same time it is the eating place and residence of numerous small creatures.

Above all these lower floors there is the crown, the tree layer arching above everything else. The struggle for survival is particularly intense in the tree layer. If we remember that in newly-planted woodland, after several years, 5,000, 20,000 or even more small trees may be present per acre, but that in an old growth only a few hundred remain, we can assess the fierceness of this struggle. Older tree growths are, therefore, usually layered by a main growth of dominant trees, an intermediate tree growth striving continually towards the top, and often a distinct, shorter subsidiary growth. These vanquished trees continue to linger on as shade trees, if Nature equips them to be content with little light. In that event, they perform important tasks: cover the ground, fill out the air space and provide protection from the wind. With their large, shade-giving leaves, spreading foliage and specialized forms of leaf and needle, they try to catch as much as they can of the weak light penetrating through to them. The others, the trees which need more light but find themselves under the umbrella of the main growth, are defeated in the struggle. In the uncultivated forest, might is right;

the spreading tree suppresses the tree with a small crown, the quick-growing overtakes the slow-growing. In the economy of the forest the struggle is weighted to the advantage of the valuable, well-formed and beautiful tree.

The living earth

The most multifarious life and the greatest secrets of the forest community are hidden in the upper regions of the rooted soil. In the basement of the forest house there prevails a busy activity and an abundance of life forms whose interaction is beyond our comprehension. Uncountable hordes of labourers work the upper soil layer. Mice, worms, millepedes, isopods, snails, larvae, spiders, springtails, wheel-animals, branchiopods and cirripedla burrow through the soil down to its deeper layers, breaking up organic waste material and mixing it with mineral soil. Earth worms pull leaves and grasses into their tubes and digest them together with mineral substances, and by their passages and holes they ensure good aeration of the soil. Many people have estimated that there are $\frac{1}{4}$–1 — others say 1–3 — tons of earth worms per acre. Under favourable conditions they produce 40–60 tons of humus a year.

All broken-up waste material is decomposed by an invisible army of the tiniest creatures. Ten milliard microscopically small animals and plants work and live in 60 cubic inches of good forest soil. Bacteria, fission-fungi and saprophytes, mites, diatoms and green algae break up the organic waste and recreate the initial substances. Many fungi and bacteria are able to dissolve compounds containing protein; others break up lignin, cellulose, tannin and fats. One small group specializes in binding nitrogen from the air in the soil. The complex picture of these soil creatures is to a large degree known, but in spite of this, our knowledge is quite modest in relation to the many questions of soil

55

biology that are still unsolved. Thousands of bacteria strains have not as yet been investigated and classified.

The active break-down of organic substances produces the brown-black, wholesome humus which, together with the clay particles of the soil, holds together the most important nutrient ions in the soil. The humus has a beneficial effect upon the water, air and nutrient balance, and creates the optimum conditions for plant growth. The world of the living soil indeed accomplishes a prodigious task when we remember that the residue of dry branches, needles, foliage, and dead creatures is substantial. Various reports claim that litter accumulates at a rate of nearly 30 cwt per acre a year in temperate zone forest, 10 cwt in the silver birch forests of the Nordic areas and 5 cwt in slow-growing mountain forests. The loss of large sections of micro-fauna and micro-flora occurs as a result of deforestation, the selfish growing of tree species alien to the locality — in particular the growing solely of conifers — and because of harmful pesticides, which restrict soil activity.

The forest as the home of plants and animals

The number of plant and animal species related to one another in the forest is greater than is generally realized. Years of joint biocoenotic research by M. Frei-Sulzer and a number of specialists has shown that in the beech forests of large areas of Germany, Austria and Switzerland there exist some 11,000 animal and plant species that are dependent on this habitat, or are attached to it directly or indirectly during one phase or another of their development.

The great expanse of the forest habitat and the profusion of living conditions make it possible for the greatest diversity of life forms to flourish. The more

colourful the flora, the richer the fauna. In the soil-cover alone live innumerable tiny animals, seeking their nourishment, protection and shelter. Each dead leaf, each pad of moss, has its importance, representing to these teeming creatures their world, as important to them as the great wide world is to us. The insects that live in the soil, in the soil-cover or in the trunk and crown areas of the forest are particularly numerous, colourful and multiform. They make up four-fifths of all animal species. They are among the oldest inhabitants of our earth, since they appeared about 40 million years before man, and their incredible power of reproduction and amazing adaptability have helped them to colonize almost all regions of land and fresh water. More than 700,000 species have already been identified and described, and innumerable species are still unknown. About ten years ago the ability of insects to make aromatic secretions was established. These pheromones play an important part in all aspects of co-existence: they convey orientation and warning signals, they serve to intimidate enemies or to attract the sexes. There are numerous forest pests among the insects. About 500 species of beetle are known that can endanger the forest to an alarming extent. In the healthy forest, not only is their mass-production prevented but they are held in biological balance and kept within bounds by an army of insect-eaters, robbers and parasites.

Another population is formed by the birds, who are especially abundant in mixed deciduous forests. They find shelter and build their nests in the shrub layers, the trunk region or the crown area. Their song can mean anything: marking out their territory, joy and pleasure, warning calls or abuse of intruders. Many birds play, as insect-eaters, an important role in keeping down forest pests. In West European forests about eighty bird species breed, many of which, however, feed in the open fields. Infinitely greater are the number, the multiplicity of form and the abundance of colour in the bird world of tropical forests.

Forest life is unimaginable without small creatures and game. To them the forest means home and food, or perhaps just shelter when they venture into the open fields and must seek safety in flight. Many animals form their own health protection service. In the densely populated countries the large beasts of prey have disappeared, considerably disturbing the natural balance in the animal world. In such countries only fox and lynx still provide natural selection among the game and prevent epidemics. Dead birds, frogs and toads are consumed by moles. Beasts and birds of prey keep down the number of mice. An important protective role is played by the red wood ant.

If in the solitude of the forest on a spring or summer day we quietly observe, we are thrilled by the incredible multiplicity of flora and fauna that live, grow and find protection there, accomplishing the tasks and expectations of growth and creation. The forest community needs all these creatures to carry on its magnificent teamwork. Nature has consequently forgotten none of her creations and has equipped them with all that they need to live, to fight or to flee.

The forest fights

To the very limits of its being, the forest conducts an everlasting battle against the disadvantages of location, against frost and cold, drought and starvation, against the sand of the desert or cascading avalanches of snow and stone. Nature has equipped the forest with the proper means to fight this battle. It is infinitely inventive where the maintenance of life is concerned. Under extreme conditions the forest grows little, lacks nourishment and is poorly managed, but in spite of this it fulfils important tasks, forms a protective layer and protects the soil from being washed away and from destruction. As well as all this, a rich variety of flora and fauna prevails here.

The mountain forest. In all continents there is a limit to the altitude at which woodland can survive. The growing period, when there is water and warmth for growth, blossoming and ripening, is short. Ellenberg asserts that the upper timber line lies at a level where an average temperature of $+5°C$ prevails for at least 100 days in the year. It is the most remarkable and most impressive vegetation limit. In the European Alps it lies at 6,000–7,250 feet above sea-level, varying with local factors; in the Southern Alps in fact larch trees are still to be found at a height of 8,850 feet. Along this boundary there is a battlefield approximately 330 feet wide where the forest becomes more and more sparse. Above this strip individual trees or small groups, mostly of gnarled dwarf types, climb still higher, but only in sheltered spots. Along this natural boundary the forest carries on a stern struggle for its existence. Here the laws are fiercer and the temperature differences greater. Along this front-line is conducted the defence against the attack of Nature's enemy forces into the closed belt of forest and into the habitat of man lower down in the valley. It is a tireless wrestle against a strong opponent, against avalanches and rock falls, thunderstorms and rain, gales and biting cold, a battle of incredible ferocity, without mercy, never-ending, a stubborn resistance to danger and distress. Many trees stand exposed as lone fighters in the foremost outposts, often misshapen, with trunks disfigured by rock falls, crowns rent by storm and snow, roots stunted among barren rocks; and yet somehow they still put out green branches, blossom and bear fruit to preserve life in this arid spot. Every tree is an obstinate character scarred by battle. Other trees have joined into groups for united resistance. Along this front there are advances and retreats, destruction and suppression, life struggling against natural catastrophes, and a tentative resurgence and advance of the forest in calmer periods. "It is a very direct, concrete and to some extent personal service that trees perform for man" (E. A. Rossmässler, 1863).

The ecology of the trees of the mountain forests is adapted to the hard environment. In the course of millennia of natural selection, climate-resistant strains have developed. Whatever failed to fit in with Nature's stern laws was rejected. Long before the end of the growing period, the trees turn their shoots into wood so that they will not be destroyed by an early winter. Not until late spring do they germinate, when the risk of hard frosts is gone. From early February to the end of April the frost brings the danger of drought, when on sunny days and with drying winds evaporation occurs in spite of the regulating mechanism of the stomata, and the roots cannot draw substitute water from the still-frozen soil.

The forest in the cold North. The forest has similar problems of existence at the sub-Arctic timber line. Here also the forest becomes sparse, more monotonous, more wretched. Only a few tree species such as conifers, the frugal aspen, birch and willow survive the long non-growing period, becoming bent and gnarled. The soil is frozen rock-hard during every winter. For four or five months of the year the trees are completely frozen from root to bud. However, surprisingly, with each new spring life awakes. The upper layer of soil, where only the smallest amounts of nutriment are available, thaws out for the few summer months. Farther North, a zone of dwarf shrubs merges into grass heath, stretches of moss and modest pulvilliform plants, until at last the ice calls a halt. A spruce species grows in Alaska under the severest conditions; in the course of a century it forms a small trunk 10 feet high and with a diameter of only one inch. The cold is not the worst enemy to the growth of trees. Trees do not freeze to death. Laboratory tests show that last year's needles of the Swiss stone-pine and common fir withstand temperatures as low as $-50°C$, and those of the red pine temperatures as low as $-40°C$. In Siberian forest regions the local spruce can even survive temperatures as low as −60°C. In spite of the marshy soil, very little water is available to the plants because the sub-soil is permanently frozen. This is why on sunny days these trees are endangered by frost-drought, although the hardy outer skin of the needles greatly reduces evaporation. In the wide, almost wholly flat areas melted water and rain cannot flow away but gathers in hollows, forming extensive marshes and vast peat layers. In these cold, wet, acid soils, fungi play the most important role in decomposing the organic plant residues. They weave through the top soil layers with intertwined threads. One square yard of Podolia soil permeated with needles may contain more than 60,000 miles of fungus threads. By obstinately continuing to grow, the forest tries to assert itself. Regeneration is difficult in spite of ample seed production, because by far the greater part of the seedlings are destroyed.

The forest on the edge of the desert. Apart from the natural deserts where Nature itself makes it impossible for the forest to thrive, there are vast border zones of steppe- and savannah-like arid regions and low-growing parched forests. Here the scanty water supply becomes a matter of life and death. The trees develop long strings of roots and relatively weak growth above ground, which can survive on the little water available. Umbrella trees and cashews often have roots reaching to a depth of 100 ft. To restrict the loss of water through evaporation, trees in arid regions have small leaves with a thick top skin and leaf pores that can close completely. In extremely dry species the leaves have degenerated into thorns, which reduce evaporation still further. Other species, such as the African monkey-bread, tree-like euphorbia or cactus trees, store substantial amounts of water in their roots or trunk for periods of shortage. In tropical regions that are dry in summertime the trees shed their leaves at the beginning of this period, so that for months the forests stand bare. However, as soon as the rains start they grow green in a very short time. It is marvellous how the ecology of the trees and of other flora is adapted to these severe conditions.

Natural re-afforestation

Thanks to its ecological vitality, the forest is always ready and able to reconquer cultivatable soil. It can be seen everywhere how open country left to its own devices sooner or later becomes naturally overgrown with woodland flora. On barren soil and in cold areas this process takes place only very slowly, but in warm, humid countries the forest reappears very quickly. Natural re-afforestation also follows its own laws. The first and usually less sensitive tree species that rush in to cover the ground are followed by bacteria, fungi, accompanying flora and fauna — indeed all that is needed to develop and prepare the ground for the more demanding species that come later — by the same process that occurred during the colonization of the raw soil in the post-glacial period. Natural re-afforestation is restricted only by time, since it is not an initial foothold but the stages of regression that make substantially quicker re-colonization possible.

Man as the enemy
of the forest

Since he came into being, man has encroached upon the surrounding landscape, utilized it and shaped it. Interference was at first modest and scarcely perceptible compared with the overwhelming forces of Nature. With increasing population and higher economic demands, these encroachments have gradually become more serious. During the post-glacial period, most land areas — where there were no swamps or other boundaries — were covered with extensive jungle. Changing a large part of these woodland areas into arable land (or simply destroying them) was perhaps the most serious of the intrusions into the primeval landscape. Wherever man settled, the forest had to yield. There are many types of human influence upon the forests of the earth, from the primitive form of utilization through manifestly agricultural forms to complete forest destruction.

The conversion of forests into arable land in Western Europe

Primitive, nomadic man in Stone Age times worked as a hunter, a hoarder and a fisherman. He existed on fungi, berries, wild honey and the wild fruit of the forest, on various herbs, on fish and on the flesh of wild animals. He needed wood for the fire in his hearth and for his tools. The forest was to him a home and at the same time a habitat. But the trackless forest was also his enemy. Beasts of prey lived there, which he tried to kill but whose prey he could become. It was not until the Neolithic Age (4,000–2,000 years BC) that man became less peripatetic, settled in village communities and began to breed cattle, to cultivate the soil and to engage in various crafts. Increasingly, areas of forest were levelled to provide space for settlement, grazing and cultivation. The most effective means of pushing back the forest was to clear it by fire. Forest fires, once started, usually continued to destroy unhindered until they were arrested by swamps or until rain extinguished them. Man was becoming the enemy of the

forest. In spite of this, in Roman times three-quarters of the land area was still covered by forest. At the start of the Great Migration in the fourth century AD, a wave of people flowed into the still extensive forest lands, forming a second mighty wave of colonization and forest clearance, whose apogee was reached in Carolingian times during the eighth and ninth centuries. It was primarily the monasteries that, from about the year 900 AD, in association with the foundation of many churches and villages, developed wide-ranging clearance activities. The introduction of the three-field system, which required significantly larger land areas, also resulted in much deforestation. In addition, the development of wine-growing in the warmer regions claimed large forest areas.

The prehistoric, ancient and medieval settlements were usually situated in the same locations — in the fertile plains suitable for agriculture, in the low-lying hilly regions and on the terraced hill slopes with their light soil. On the other hand, swamps and valleys threatened by flood were avoided, as also initially were the rugged, mountainous areas. Medieval woodland cleared for farming transformed more than half the forest areas into meadow and arable land. By the fourteenth century — in some areas by the fifteenth — the main conversion period had come to an end. The division between open cultivated land and forest had been largely concluded. Since farmers had mainly taken over those areas more suitable for cultivation, the forest was left in the less suitable locations, such as north-facing hillsides, steep mountain slopes, mountain peaks, gorges and regions with arid and shallow soil. However, the process has continued up to this very day, for all over the world large or small areas of forest are continually being converted into arable land. Even in this century, in most European countries further important forest areas are unfortunately disappearing as a consequence of increasing urbanization, high-rise buildings and the construction of roads and industrial installations. Forest-clearance did not come to a halt at the foot of the Alps. Even at the time of the Celts, in the last century BC, colonization penetrated into the most remote mountain valleys. The Alemanni established countless new farms and cleared vast forest areas for agriculture. The massive timber supplies required by the Alpine economy for the growing villages, for the many crafts and for the construction of roads and tunnels through the mountain passes gradually led to entire valleys being deforested. As a result of this, the natural upper timber line in all accessible locations was probably lowered by 650–1,000 ft.

Forest destruction overseas

It is not a difficult process to destroy the forest completely. For very many years man has done this in the crudest possible form. Over-population in Asia resulted in the destruction of forestland over enormous areas in the course of more than 2,000 years. In contrast to this, rapid deforestation for capitalistic and industrial reasons came to wide areas of the USA, where woodland in the East, in the Mississippi region and in the Middle West was exploited by the gigantic timber industry. In the same way as mines and other available raw materials are worked without consideration or forethought, the timber from extensive primeval forests that once seemed to be inexhaustible was plundered without a thought being given to its renewal.

In most areas of the world the intensive grazing of sheep and goats brought on the destruction of the forest. Forests used as grazing areas, in which natural renewal is continually destroyed, become old, thin out and finally disappear. In an even more brutal and direct way, clearance by fire has destroyed the forest. The nomad especially has in tropical countries burned away wide forest areas in unbroken succession to employ them, sometimes using intensive methods, for growing millet, maize, rice, manioc, bananas, sugar cane, cocoa, coffee, oil plants and spices. In the tropical rain forest region in particular, impoverishment of the upper soil layers occurs very rapidly by the mineral salts being washed away, so that agricultural exploitation is usually possible for only two or three years. The farmers leave these areas and carry out new clearances with the help of fire. The barren soil is left to its own devices. It takes ten, twenty or even thirty years before these areas again grow into a secondary forest with shrubs and trees, but such a forest will be lower in growth and poorer in species than was the original virgin forest. Frequently this shrubland is cleared again after twenty or thirty years, thus accelerating the total destruction of the soil. The population explosion is at present growing at an alarming pace in the developing countries, most of which are located in the tropical belt, and there is increasing pressure on these nations to make agricultural land available by burning woodland. Nomadic cultivation is carried on by more than 200 million people in Africa, Asia and Latin America, resulting in the destruction of more than 500 million acres of forest a year.

Between the tropics of Cancer and Capricorn there are about 4.25 milliard acres of forest, or 45% of the world's woodland. These areas include the secondary forests caused by fire and hoe-cultivation, and other forest communities destroyed to a greater or lesser extent. In the natural tropical forests there may be nearly 300 different tree species on just two acres. But in fact there is only 45 cu ft of usable timber in such an area. Therefore, to secure large yields of valuable timber still more vast areas of rain forest are devastated. FAO experts have estimated that in the period 1954–1968 good, fully-stocked rain forest was lost through excessive timber exploitation to the extent of 37 million acres on the West Coast of Africa, 22 million acres in South-East Asia and 39 million acres in Latin America. These figures do not include deforestation to meet domestic needs, nor the areas burnt for nomadic hoe-cultivation. Since 3,000 million cu ft of timber was felled in the tropical zones of Africa, Asia and South America in 1968, experts predict that the tropical forest will be completely destroyed by the end of this century. The estimated timber exports for all three continents up to 1985 amount to 3,300 million cu ft. In view of the quantity extracted per acre by the production methods now employed, this means that 325 million acres of forest will be devastated. Added to this is the amount destroyed for domestic consumption and clearances by fire for arable land. The total loss of productive forest by 1985 is estimated to be 500 million acres for West Africa and South-East Asia alone. In Latin America losses may be even greater, and it appears not impossible that "at the present rate of destruction even the gigantic forests of the Amazon basin will fall victim to the axe and fire by 1985, except for small remnants in the reservations" (P. Sartorius, 1971).

In the tropics, there are today on the one hand forests completely destroyed by fire and plundered by the thoughtless felling of valuable timber, and on the other hand wholly unexploited and unexplored virgin forest that represents the world's last timber reserves.

The natural forest never dies. No catastrophe, not even fungoid diseases or the immoderate use of pesticides, can permanently destroy a forest. Again and again, from residual trees or from neighbouring areas, seedlings appear, new trees grow, and from the transitory regression stage a forest is sooner or later recreated. For millions of years the virgin forest has persevered, has become too old and has been destroyed in some areas, but it has always regenerated itself. The forest is the most stable and most powerful of all plant communities, and it does not voluntarily yield up an inch of ground. It is man alone who finally destroys great areas of forest. Not only this, but he has changed the landscape to an extent that is scarcely believable. All the original landscape has disappeared; in every country the forest has been cut back to a fraction of its original extent.

Changes in the forest's structure

While the woodland was being gradually reduced by clearance, the remaining forests were also being increasingly exploited. Although this affected the composition of the forest, the natural pattern was maintained for a long time, because the regenerative powers of the forest survived these encroachments. It was not until man began to cultivate forests, in particular by the planned planting of seeds or seedlings, that the original pattern of forests was finally destroyed. The lower plains of Central Europe were covered naturally by mixed forests, and in extensive areas entirely by deciduous trees. This primeval pattern was abruptly changed, especially during the wholesale deforestation period dating from the beginning of the nineteenth century, by the selfish desire to produce coniferous trees, which are economically more valuable. The anthropogenic effect became the most crucial factor in forest development. In the common struggle between tree species in the growing forest, man's priorities differ from those of Nature. The ecological balance was disturbed by growing purely artificial stands covering large areas. In the artificial forests there was a permanent deterioration of the soil and an impoverishment of the flora and fauna. In particular, the problem of pesticides became sharply evident. In many areas these chemical poisons were used in an attempt to restore the ecological balance, thus weakening the defensive forces inherent in the environment. Nature has protested against this unnatural phenomenon by violent calamities, and repeatedly destroyed forest stands which are alien to their environment by storm, insect attack or fungoid diseases.

It is of great importance that the accompanying flora of the original natural forests has changed very little in spite of artificial planting. Ground vegetation is amazingly faithful to its habitat, and lingers on in the top layer even when conditions are completely changed. Certainly the biologically weaker species become eliminated, but representatives of other communities appear in increasing numbers. In spite of this, the accompanying flora displays its original composition, the most important and characteristic species clinging stubbornly to their natural habitat. By this means it becomes possible for the expert to identify the type of natural forest community that would be present in the location concerned but for the interference of man.

Artificial stands cannot continue to exist without constant husbandry. In their interior a quiet, tenacious invasion can be observed. From the edges of the forest, from the roads and from any gaps in the stands the original species of tree and shrub gradually begin to penetrate and reconquer their hereditary habitat. Thus a specific stock colonizes the sub-layer and with uninterrupted vitality perseveres in a gradual return to the natural stand. The outcome of this process can in no way be doubted, provided man does not again interfere and, by his desire for profit, once more force events to take a different course.

Life without the forest

The longer a country is populated, the poorer it becomes in forest and water. The Alps and the tropics provide startling examples of how Nature takes revenge with a ruthless lack of mercy for the destruction of its forests. The chain of events is disastrous. The stripped forest soil becomes compressed; instead of soaking into the soil, most of the rain water runs away on the surface. The previously balanced water economy of the rivers is fundamentally and permanently damaged. Fertile soil is washed away, slopes subside and the movement of sedimentary material increases. The rush of water flowing downwards becomes a torrent, which in its lower courses floods large areas and devastates them with waste matter and soil. Once soil erosion has started it continues at a frightening pace. Shortly after the rainy season the rivers dry out, the springs are exhausted and the ground water-level falls. The consequences are shortage of water, drought, land turned into steppe and destroyed, and long-term deterioration of the climate. Too much water causes an emergency, but so does too little water. The heat of the day and the cold of the night work on the bare rock and naked soil. Rock and soil disintegrate and are ground to dust. In strong winds, soil particles of one fiftieth of an inch diameter or less are blown away. A tropical hurricane can, within a few hours, carry away a soil layer of several inches thickness. We should never forget the slopes in the Mediterranean region mutilated by erosion, the dust deserts in the American Middle West, the tremendous struggle against soil destruction by wind and water in the Mississippi area, and the incessant encroachment by the African sand seas. Forest destruction has had critical economic and social consequences. In China, India and North Africa it has caused the poverty and famine that have befallen the peoples of these old civilizations. The waste areas of the world extend over nearly 8 million square miles. Where the forest has disappeared, man has lost, deservedly, the necessities for sustaining life.

Where the forest dies, the desert grows. There are of course natural deserts, where the lack of rain, naked rock, ice and frost oppose the prosperity of a vegetation cover. But the desert areas caused by man are even greater. The wide savannahs and the treeless steppes are, many scientists believe, less a natural climatic formation than, very possibly, merely a stage in the degradation of former forest communities, which in the final phase will be turned into complete desert by grazing and destruction of the last sparse vegetation cover.

In 1940 the World Forestry Commission considered eight areas:

Europe: densely populated; forests carefully husbanded.
Soviet Union: moderately populated; forests as yet little husbanded.
Middle East and North Africa: very densely populated; over-exploited forests.
North America: moderately populated; forest husbandry impending.
Central and South America: moderately populated; extensive, still unhusbanded tropical forests.
Africa (less North Africa): moderately populated; extensive, still unhusbanded tropical forests.
South and East Asia: very densely populated; insignificant forest areas.
Australia and islands: sparsely populated; limited forest areas; husbandry at the initial stage.

According to the second FAO World Forestry Survey published in 1955, almost 30% of the land areas of the world – that is, about 15 million square miles – were afforested; of this, only about 30% was economically exploited.

Recent investigation shows that forests may be classified as follows:

1 million square miles of intensively husbanded forest;
$2\frac{3}{4}$ million square miles of barren areas, no production, reafforestation areas;

4 million square miles are exploited with no consideration of the future;
8 million square miles of virgin forest.

Mediterranean region and Near East

The Iberian peninsula was formerly dressed in a gown of evergreen; in Spain today only 8% of the land is still woodland. Deforestation has also been very heavy in Italy. "No other country displays more impressive proof of the dreadful effect of deforestation in impoverishing the soil and damaging the water economy" (Barbe Baker). Before the great migrations, there was still dense forest in Istria and Dalmatia. At the height of the Byzantine period in Venice and Genoa, there was an enormous demand for timber for the construction of cities and of mercantile and war fleets. The timber supplies came from the oak forests of Dalmatia and Italy. In the fifteenth century the city of Venice possessed 3,500 ships.

Great forests were destroyed by fire, but above all by the unrestricted grazing of goats and sheep. The intensive felling increased the erosive power of water on the steep slopes. Summer drought and grazing permitted no new forest to appear. The green grazing land became barren and unproductive; about 1,500 square miles became limestone desert. "I know of no other place on earth so reminiscent of a moon landscape as these karst hills. There are only two colours: the blue of the sky and the sea and the blinding, scorching white of the rock" (R. H. France).

The Eastern Mediterranean region was also formerly covered in extensive forest. In the course of a long development period, by war, fire, over-exploitation and primitive grazing, the belt of forest, once 50 miles wide, extending from Asia Minor to Gaza, became a landscape destroyed by erosion, a desert of rock and mountain. Today in the Lebanon there exist only a few hundred trees representing the scant remnant of the once-famous cedar forests. Once 65–70% of Greece was covered in forest; nowadays the forest covers only 15% of the land area. Extensive regions became rough, stony and barren in spite of a climate so favourable to the growth of woodland.

North America

The United States presents an impressive picture of a comparatively recent, and therefore all the more harsh, destruction of forest to the greatest extent possible. It is the most striking example in civilization's brutal history. As recently as 200 years ago, only a small proportion of the present-day states had been colonized. Vast areas of the huge land strip between the Western and Eastern mountain chains were covered in forest. The supply of timber from this virgin forest seemed inexhaustible. Exploitation steadily advanced westwards from the East coast, gradually moving inland until the forest stands of the Rocky Mountains were reached. "Treeless, burned-off, exhausted by excessive grazing, stripped of humus, unfertilized and uncared-for, the soil was overtaken by its inevitable fate." Even when it reached extensive areas that naturally tended to turn into steppe, deforestation did not stop. From being a former timber-exporting country, which until 1909 was able to meet the demand for pulpwood from its own resources, the USA became the largest timber-importing country in the world. It is estimated that $1\frac{1}{4}$ million square miles of forest still existed 300 years ago, covering almost half the ground area. The forest has now shrunk to a quarter – in the Mississippi region to less than one tenth – of its original extent. During the period 1780–1930 alone, in 150 years, over half a million square miles of forest was cleared for farming. The naked soil was rapidly eroded by water and wind. To date, about $1\frac{1}{2}$ million square miles of land has become barren. It is estimated that cultivation is no longer possible on one-seventh of the total ground area of the USA. In 1934, three milliard tons of fertile humus in the American Middle West were blown nearly 2,000 miles into the Atlantic by a hurricane; 170,000 square miles of arable land were destroyed, and 210,000 square miles were severely damaged. In its estuary area the Mississippi has pushed its delta more than sixty miles into the sea within a century.

At the United Nations Conference in 1963, scientists declared that the most recent investigations showed that $2\frac{1}{2}$ million square miles, about half of all the cultivated land on this continent, displays erosion damage. Not until the presidency of Theodore Roosevelt was a forestry service established. In a message to Congress shortly before his death, Roosevelt said: "Many millions of acres of ground must be replanted with grass and trees if we wish to prevent the creation of a new man-made Sahara."

Central America

Charcoal-burning, forest grazing, clearance by fire and nomadic hoeing has also led to enormous forest destruction in this part of the world. Pressure from an expanding population has had the effect of continuously shortening the fallow period, when the bush could grow again in abandoned areas. Stands on the timber line, that were in no way suitable for agriculture, have been cleared. Forests up to a height of 11,500 ft have been up-rooted for maize growing, which is possible only up to a height of 10,000 ft. The consequences are soil erosion, dust storms and lowering of the ground water-level.

South America

With South America's high rainfall, the great destruction of the forests was bound to lead to catastrophic soil erosion. According to an FAO report, 20,000–40,000 square miles are under clearance in Latin America to provide soil to feed the rapidly growing population and raw material for the timber industry and for export. In the Parana river area alone, more than 20,000 square miles of forest were cleared within twenty years. These areas are used for grazing, and will soon be completely barren. "In no part of the world have so few destroyed so much" (F. Erico). Particularly intensive was the forest destruction along the entire West Coast of the Andes. After felling the useful timber, the remnant was set on fire to provide cattle pastures and to exploit the region for agriculture, until the soil was destroyed by erosion or evaporation. It is said that in Chile 60,000 square miles, or almost one-third of the country, was covered with forest. Today there are only 7,000 square miles of forest with some commercial value. The Alerce pine, which grows up to 330 ft in height and to an age of more than 2,000 years, has disappeared from Chile. In association with deforestation, deep furrows were cut into the slopes, and the valley beds became filled with rubble. In the tropical highlands of Bolivia, A. Heim observed extensive burnt-off slopes that had been planted with cocoa bushes, "If the forest soil becomes exhausted after four years, adjacent areas are cleared and the abandoned slopes are left to erode." In Brazil, coffee cultivation has brought about the destruction of large parts of the land. The forest was cleared and the soil exposed to erosion. The latest government project is to provide arable land by extensive clearances in the 3,000 mile-long Trans-Amazon area in order to settle thousands of families. It may be anticipated that this accelerated development will soon lead to exploitation of the Amazon basin, the largest virgin forest area in the world.

Africa

Many scientists have come to the conclusion that North Africa's present forests are only a small remnant of what were once extensive forest areas. Some hundreds of years before Christ, Carthage, on the border of present-day Tunisia, had 700,000 inhabitants. Tunisia was the granary of the Roman Empire. Expeditions that began to explore the North African desert in the second half of the nineteenth century found beneath the desert sand traces of former cities, fertile soil and art treasures that bore witness to the advanced culture of the people who once inhabited these lands. Forest destruction had already begun in ancient times. Widespread cultivation and unrestricted cattle grazing devastated the forest stand, crucial for climate and water economy, leading to the disappearance of springs and wells and to the fall of the civilized states. Of Ethiopia's original forest area of some 175,000 square miles, 140,000 square miles, or 80% is completely, and probably permanently, destroyed (W. Bosshard, 1961). The unavoidable consequences are erosion, the slipping away of entire slopes, unfavourable climatic changes, impoverishment of agricultural land and flooding. The African rain forest has had to forfeit at least 400,000 square miles of its former expanse. In the period 1930–1970 alone, between a quarter and a third of the original rain forest – in areas near the coast, more than a half – was destroyed (Institute for World Forestry). In Nigeria, where at present 77 square miles a year are cleared just for the cultivation of mountain rice, the virgin forest was reduced from 38,000 to 13,500 square miles within a few decades. It is thought that by 1980 only one-fifth of the original forest will still exist. In the Congo also, the rain forest that once covered 80% of the country has shrunk to less than half its original area. Cameroon is reported to have granted concessions in 1969 and 1970 for the felling of about 20,000 square miles of a recorded forest area of 95,000 square miles, of which only 38,000 square miles still stands untouched. On the Ivory Coast, one-third of the recorded forest area fell victim to nomads who burned the forest between 1956 and 1966. "It may be anticipated that in the West African rain forest all woodland outside the forest reservations will have been transformed into bush, plantation and savannah by the end of the century." This means that within 25–30 years about 45% of the virgin forest currently in existence will have disappeared.

On the Ivory Coast, the great desert advances by half a mile every year. West Africa's deciduous forests, including savannah and bush, cover 2 million square miles. Between 1965 and 1985, 300,000–400,000 square miles will be required for export, domestic needs and fire clearance, "so that the exhaustion of these forests in the foreseeable future can scarcely be checked" (P. Sartorius, 1963).

Asia

Due to special climatic, edaphic and topographic conditions, extensive parts of the Asiatic Continent have always been poor in vegetation and populated to only a limited extent. Even more devastating has been the effect upon the natural environment in densely populated regions. The homes of these ancient civilizations have over the last thousand years become almost completely deforested and rendered barren by drought. The further East one goes, the more staggering is the picture.

Asia Minor was once a land covered by forest. The Roman Empire's provinces richest in woodland were located in the great basin of Inner Anatolia. Today these are arid, parched areas of steppe and salt lakes, inhabited only by shepherds and sheep.

The great empires of Babylon, Syria and Persia declined as forest clearance increased. "The desert crept nearer." These deserts tell of the misery of an abused, worn-out soil. India became the cradle of civilization when the land was still largely covered by forest. The forest today accounts for only 18% of the land area. China also used to be a country rich in forest. It is one of the countries with the oldest agricultural heritage, and today only about 9% is forest. Vast areas of this gigantic country are among the most eroded landscapes in the world. Millions of square miles of fertile land have become deserted areas of sand, mud and gravel. In the Far East, according to an FAO report, 32,000 square miles of forest a year is destroyed by about 24 million nomadic forest-burners. One-sixth of the land is in danger of erosion. Each year the Yellow River washes 2.5 milliard tons of soil into the sea and the Yangtze, 0.6 milliard tons. The endless deserts in the upper stretches of these rivers are among the most over-exploited and wind-eroded areas in the world.

Each year more than 60,000 square miles of forest is lost in clearance in South and South-East Asia. Indonesia leads with 20,000 square miles followed by India with 19,000 square miles and Burma with 4,500 square miles. In Thailand, where 15,000 square miles of forest has been destroyed in recent times, it is planned to reduce the present woodland area of 110,000 square miles by a further 39,000 square miles for agricultural purposes. It is estimated that deforestation in this area will cause about 300,000 square miles of land to become barren and to degenerate into steppe and bush.

The percentage of untouched forest in South-East Asia amounts to only some 25–35% of the total forest area of $1\frac{1}{2}$ million square miles; that is, 400,000–500,000 square miles. The loss of woodland between 1965 and 1985 is estimated at 325,000–475,000 square miles, so that here, too, the exploitable forest will be exhausted by the end of the century.

Australia

In Australia, large areas are naturally bare and arid. To some extent the Australian desert can, however, be attributed to deforestation. The first inhabitants protected the forest, since their demand for wood was slight, but when the new settlers came forest destruction began. Without a thought for the future, great areas of forest were plundered. Entire mountain regions were robbed of their greenery. Particularly desperate was the senseless destruction of the extensive and beautiful eucalyptus forests. Today, thousands – indeed tens of thousands – of yellow cracks in the sheep pastures on the bare mountains bear witness to the steady increase in soil destruction.

The tragedy of forest destruction

Of the world's land area 34% is covered by forest. 41% is barren and unprofitable land. Most heavily afforested are South America (47%), Canada (34%), the USA (33%) and the USSR (33%). The percentage of woodland in Europe is 28%, in Central America 27%, in Africa 27%, in Asia 19% and in the Pacific area 14%. It must however be borne in mind that very extensive areas are merely unproductive, sparse, non-profitable woodland. Acres per inhabitant are in Canada 59, in South America 18, in the Pacific areas 16, in Finland and the USSR 13.5, in Africa 8.25, in the USA 4.25 and in Western Europe less than three quarters. How much woodland has disappeared over the whole world is impossible to say.

One might believe that these shattering consequences of deforestation would have made people come to their senses and take the most stringent measures to put a stop to further devastation. But more and more people came on the scene who did not know the earlier circumstances and who continued deforestation without thinking about the inevitable outcome of the process. Objective observation of Nature and appreciation of its cohesion were lacking among primitive peoples. The annual cycles of growth and withering, rain and drought, flood and famine were accepted as the work of supernatural forces and were tolerated as part of an unavoidable fate. It is a tragedy for the entire world and for all peoples that forest destruction has gone on for so many centuries until this very day without man having been aware until recently of what he was doing. The necessary insight to apprehend the acceleration of this destruction has been severely limited, and one generation is too short a period to realize the consequences in their entirety. If we make an effort to reflect upon these processes taking place over vast areas and over long periods, it is horrifying that these crimes committed against the body of our earth are still being perpetrated. Over a long period man has transformed what were once fertile areas into barren land. Only one-tenth of the land surface is at present usuable for food production, while in all parts of the world appalling exploitation has taken away from millions of square miles their ability to feed mankind.

Warning voices have been raised all over the world for many decades. Large organisations have been formed to conserve the forest. They opposed the reckless exploitation of the soil, but all too long ago they were condemned to failure. Many people have today been roused by the ever-worsening reports. But the time for self-denial is not yet ripe; there is still a gap between the necessity to observe moderation and the urge to increase our standard of living.

The forest as the friend of man

From time immemorial there has been conflict between man's ruthless requisitioning of land and the struggle for self-preservation by forest and Nature. The result of this conflict was usually to the disadvantage of Nature. Material greed proved to be stronger. The forest's history is linked with the cultural development of great nations, and it differs, therefore, between countries. "But one thing is certain all over the world: wherever man appears he brings with him his human problems. Nature is creation, living and passing away; it knows no guilt or atonement. But problems arise and grow to monstrous proportions in the Nature/man dualism" (F. v. Hornstein, 1950).

For primitive man, the forest was a place of mystery, or even misfortune. It is immeasurably great, and one can lose one's way in it. The forest, therefore, is associated with superstition by all peoples. Wood spirits live in the darkness of the forest; Pan for the Greeks, Silvan for the Romans. Helpful or wicked satyrs and dryads, wild woodmen, gnomes, fauns and demons live in the forest. There are dragons in the forests, caves hiding robbers or witches, giants or dwarfs. As it still does to the child, the forest was thought to harbour all kinds of danger and mystery, all that is sinister. If we walk in a dark forest with a child, a little hand will push its way into our large hand; the child wants to be led, it seeks protection. It peers anxiously around if a twig tugs at its clothes or a mysterious rustling is unexpectedly heard. Even the adult is not entirely without such sensations as he walks with uncertain step through the sinister blackness of the forest in the darkness of the night. There are mysterious shadows behind dark bushes, an unseen animal flees away or a bird cries its warning. Then the primitive man awakens in us; then we know fear.

Certain religions consider the tree to be man's oldest kin. Deep wisdom is hidden in this ancient belief, since in fact there were trees on this earth before man appeared. Man was originally in fear of the tree. The tree had a soul; it was

116

the seat of supernatural forces. Not only could protective and healing forces dwell in trees and plants, but also evil spirits. Many tribes honoured venerable trees as deities. The tradition of the holy tree permeates all religions, cultures and mythologies. Legend, superstition and mythology determined for long periods man's behaviour towards the tree and the forest. In the wooded groves deities were worshipped; here stood the temple, the altar where sacrifices were offered. The people gathered under the trees; counsel and judgement were given here. In joy and in sorrow, the tree was turned to as a comprehending being. Many customs arose out of this attitude towards the tree. To the Japanese, the tree is a sacred being. A prayer is said before it is felled to appease its soul. Because the tree may house an evil spirit, wood-cutters in many mountain valleys cut a cross in the trunk after felling the tree to drive out the evil. Originally, cutting a cross was a mythical means of defence; today it is often employed as a test of skill. The good tree-spirits protect man. The lovers' custom of cutting hearts and initials in the bark of trees may perhaps derive from this superstition. Among the holy trees of antiquity the most important was the lime. The mighty oak was looked upon as a symbol of life. As early as the Middle Ages it was protected as a fruit-bearing, "begetting" tree. Whoever committed sacrilege against the oak was threatened with dreadful punishment under the common law.

Even today there are popular customs in many places that may go back to veneration of the tree or tree-spirits, such as men decked in foliage, deer men, Morris dancers, etc. On church holidays, rooms are decorated with green plants; in Scandinavia the May beech is set before the front door; in other areas may-poles are erected, or young trees with spreading branches. The custom often still prevails of planting a young tree as a protection or a memorial when a child is born or as a reminder of important events. It was not until much later that enlightenment and scientific development caused man's fear of the forest to retreat and, in contrast, the beauty of the forest moved him and filled him with reverent awe. It was then that man realised that the forest was becoming increasingly important to him.

Wood as raw material and building material

The forest with its fruits and beasts provided the first man with food, habitation and clothing. But as culture increased, timber for fuel and for building material became more important. The economic history of nations is inconceivable without timber. It has accompanied mankind from his beginning through all eras. All that man created was made of wood, had to do with wood. In Central Europe, villages, market places, towns, fortifications and bridges were built of wood. Timber was employed in increasing quantities for the upkeep of houses, roads, paths, water pipelines and wells. The consumption of wood for fuel increased. All farmland was surrounded by wooden fences. The widespread vineyards needed an endless number of wooden supports. By the end of the first millenium mining had commenced and with it the exploitation of the timber supplies of huge forest areas for the mining of ore. Glassworks and brickworks used enormous quantities of wood. Timber was the indispensable building material and the fuel for all trades. The population was still small, but, in spite of this, the demand for timber in medieval towns and villages was already so great that the virgin forests could scarcely meet it. The threatened lack of timber was first countered by prohibitions. There was, as yet, no thought of increasing timber production in the wild forests by husbandry. Decrees and bans, however, were not of much help. The timber emergency became greater. Not until the second half of the eighteenth century, and especially in the nineteenth century, did far-reaching changes come to Central Europe, when eventually, although initially very hesitantly, care and cultivation of the forest was begun. Until this very day, in spite of concrete, steel and manifold synthetic materials, man has found nothing that is wood's equal. Even though wood is sometimes displaced by synthetic materials, new ways of using it become apparent in other directions. Wood has in fact been rediscovered for interior decoration. When a man arrives home from his frustrating employment he wants comfort and warmth. This is why wood is increasingly used for interior decoration and furniture. Moreover, wood is increasingly being used for paper and for packaging material. All the forecasts are that in all countries the demand for timber will increase strongly. The FAO states that demand will increase more than production, which is conditioned by natural growth. There are plans for Europe to reduce by 1980 the anticipated short-fall in timber supplies by extensive re-afforestation. Since EEC statistics show that some 12 million acres of agricultural land will be surrendered during the coming decade, at least part of this fallow land must be converted into woodland.

People in the mountains

Nature has emphatically demonstrated to man that he must go into partnership with the forest if he wishes to safeguard his well-being. This knowledge came early to the inhabitants of the Alpine villages, particularly at the time of the medieval forest-clearing, when it was observed that destruction of the forest unleashed all the forces of devastation. The threat from Nature, from the river in the valley, from avalanche, rock-fall and landslide had forced the people from the beginning to form close village communities to meet emergencies. Everything here rests upon the conservation of the forest — the opportunities for settlement, the fertility of the valleys and the safety of communications. Without the forest the mountain valleys were uninhabitable. In many Swiss valleys, therefore, the forests above the village were protected by law as early as the fourteenth century: in 1365, a decree relating to a piece of woodland in Seelisberg; in 1382, a Flüelen decree referring back to an older interdict; in 1397 an Andermatt interdict. Many of these bans were apparently promulgated as last-minute measures. Thus, the Andermatt interdict dates from a time when the Urseren valley had already been deforested, except for a few remnants, and it was an attempt to salvage what remained of a forest that once covered the entire valley slope. This interdict banned all exploitation of the forest; any person contravening it was to pay five pounds to each inhabitant of the valley, and, if he could not pay, he was expelled from the valley and never allowed to return. The document was renewed in 1717 by the valley community, and in 1735 it was endorsed by the county magistrate in connection with an offence. From the idea behind the ban on cutting the forest, the "protected forest" concept was later evolved; this term was coined in Switzerland in the Forest Act of 1876 covering the mountains and Alpine foothills, and is included in the Forest Act of 1902 still in force today. However, it was the high-water catastrophes, in conjunction with extensive over-exploitation and forest devastation, increasingly severe floods and the nineteenth century water shortage, that finally forced people to realize that woodland outside the valleys also has a crucial protective role.

New forms of danger to civilization

In recent decades, people working in science, technology and industry have achieved wonders and realized dreams that seemed unattainable only a short while ago. But these great successes have all too long concealed the truth that, behind the gleaming façade of progress, an intolerable deterioration in life's

natural foundations was taking place. Increase in population, expansion of industry, rising demands for better living standards brought with them exponential demands on the total environment. "The greatest wealth lies in the lack of needs" (Günther Schwab, 1958). Freedom from need was destroyed by clever advertising.

Man's power to change substantially the nature of his surroundings has risen to almost unlimited proportions in the last twenty-five years. The struggle with Nature has become gradually more intense. Pollution and contamination of surface water, and even of the oceans, goes on without check. Unpurified sewage, fertilizers, mineral oil, highly-effective pesticides used in incredible quantities, and especially the persistent chlorinated hydrocarbons that are characterized by their toxicity and stability, endanger our drinking water. Many noxious substances that initially contaminated only surface water have increasingly seeped down to the ground water. Oil loss in transport is estimated at 0.5–1%, which means that each year about 2 million tons of mineral oil finds its way into soil and water. One gallon of oil makes one million gallons of water undrinkable. The contamination of waterways by heavy metals, mainly mercury and lead, has already resulted in serious illness and even death. Some of these decomposition-resistant toxins, which Nature itself does not know and did not invent, are preserved in the soil for years, slowly absorbed by plants and, via the food chains, passed into the bodies of animals and of man. One quarter to one third of the enormous quantities of fertilizer used in agriculture is washed away and over-fertilizes rivers and lakes. The atmosphere also deteriorates more and more. Over large built-up areas there is a constant layer of dust and haze, clouding the sky and absorbing as much as 20% of the sunlight. Most important of all, below this layer of haze, there prevails a lack of ultra-violet light; rickets, tuberculosis and cancer are on the increase. In the last ten years the average amount of sunlight over Paris has decreased by 25%; misty days rose from 80 to 150 a year. In densely populated areas, inconceivable amounts of poisonous gases, such as carbon monoxide, sulphur dioxide, chlorine gas and fluorine gas, hydrocarbons, nitrous oxide, etc., are discharged into the atmosphere from heating plants, motor vehicles and industrial installations, and their cumulative effect makes them all the more dangerous. A particularly severe toxic fog caused almost 4,000 people to die from poisoning by the atmosphere in London in December 1952. Over the world, the annual discharge amounts to 200 million tons of carbon monoxide, 80 million tons of sulphur dioxide and 20 million tons of neutrose gases. Before men and animals become endangered, long term exposure causes the green vegetation cover to suffer physiological damage even at a low concentration of such poisonous gases. Plants near the sources of emission show the effects of acid, the forests die away. According to Linzon (1966), the area affected by smoke in the neighbourhood of three great metallurgical plants in Ohio, emitting 6,000 tons of sulphur dioxide every day, amounts to 450,000 acres of severely affected woodland and nearly a million acres of slightly affected woodland. It is cautiously estimated that the areas affected by smoke in Eastern Germany amount to 500,000 acres of woodland, and in the Erzgebirge of Lower Saxony 150,000 acres. In North Rhine-Westphalia, some 250,000 acres of forest have died. The use of resistant toxic agents to protect plants and of resistant pesticides has led to the poisoning of the soil and contamination of foodstuffs over wide areas. In spite of this, reports flow in from all parts of the world that pests feeding on our cultivated plants are on the increase. Many pathogenic agents and harmful insects have become resistant to poisons, so that even larger quantities and more effective poisons must be employed. Unfortunately, our knowledge of the undesirable side-effects of pesticides is still somewhat scanty. Until we know more about the ecology of the countryside

130

and its exposure to danger from resistant poisonous agents we have no right to continue the poison warfare at its present level, let alone to increase it.

The residues left by the use of growth-promoting agents in cultivation, of antibiotics for advancing maturity in cattle and of chemical substances for colouring and preserving foodstuffs cause injury that may appear only after a long period. "Our earth is already infected by biphenyl. The stability of this compound guarantees that it will be present for very many generations" (O. Klee). In 1970 an investigation was conducted into the herbicide used to destroy the Vietnamese jungle. This chemical displays a high dioxine content and led to the severe poisoning of soil, plants and animals. The quantities of poison spread across Vietnam would have been sufficient — if used directly — to destroy two million people. There is further risk to man. Cities are hammered by noise. Intense, continuous noise can lead to physical suffering. Man finds himself under a constant condition of stress; he is always on the alert. Noise prevents effective relaxation and disturbs beneficial sleep. Removal of the incredible amounts of waste, among which the scarcely decomposable synthetic materials continually increase, is becoming a difficult problem to solve.

In contrast to Nature's balance in the countryside, man's economy is quite unbalanced. We have continued to exhaust the soil by increasing the despoliation of green areas and by using artificial fertilizers and poisons in agriculture; we consume more and more water, increasing the burden on waterways by adding harmful foreign substances; we increasingly use up our healthy air and contaminate the atmosphere; we make ever higher demands on the forest, with consequent reduction of woodland. Above all, the economy of industrial countries is characterised by progressive consumption of raw materials of every kind. According to W. Stumm, modern man consumes 10–20 times — in industrial countries 50–100 times — as much energy as he requires for the metabolic process. The heat-loss problem associated with this may lead to consequences about which we are still completely in the dark. The problems of radio-active contamination and radiation risk to man are still unsolved and uncertain in their effects. It is indisputable that we are faced with a new threat.

For thousands of years Nature has held man in check in the mountains, along ocean coasts, in the icy Arctic and in the arid desert. Man lived in fear of the forces of Nature, which appeared to be hostile to him. In the course of time he learnt to compromise with Nature, to utilize it without substantially interfering with its course. With accelerating prosperity and improvement in technical devices, he then began a direct attack upon the soil. Restrictions on his exploitation of this apparently inexhaustible wealth were cast aside. Man today has in his hands such powerful agents that he can destroy his environment, his habitat and thus himself. Over a comparatively short period, man has in fact already caused irreparable damage to Nature. The deterioration of the natural foundations of life came about with a sinister acceleration, creating problems that are frightening in their complexity and apparent or actual hopelessness.

It is profoundly horrifying that many animal species have already been exterminated, that thousands of plant species are extinct or threatened with extinction. But many people have unlimited faith that through science and technology we shall be able to solve all problems. If, however, the world economy continues to be as careless as in the past, population growth, the exponential consumption of non-regeneratable raw materials from soil, water and atmosphere, the strain caused to the environment by dangerous substances and by the problem of waste disposal will lead uncontrollably and unavoidably to such imbalance in Nature that the ecological system must break down. Nature itself might then harshly and brutally interfere, perhaps extinguishing life on this

132

earth to begin its development once again. Man's arrogance towards his environment will not go unpunished.

"He would live a little better,
had you not given him a glimmer of heavenly light,
he calls it reason, and uses it only to be
more bestial than any beast."

(Faust, Prologue)

New tasks for the forest

Man is no longer endangered primarily by Nature. The dangers of civilization that he has himself created seem to be much more threatening. This development gave forests all over the world unexpected new uses and increased appreciation, especially those forests in the densely populated and highly industrialized areas.

The forestland's special biological-ecological capacity makes it possible for it to prevent, or at least perceptibly diminish, many of the dangers that threaten civilization. The forest is closely bound up with all the problems facing us today, and thus achieves an importance whose extent had not been realised by the turn of the century. Furthermore, we must protect the soil against further destruction. "The continued existence of man — that is to say, of his soil — is dependent upon the continued conservation of trees" (Roger Heim).

In the cultivated areas, the forest has assumed an importance that cannot be too highly assessed. Nothing disturbs man as much as a negative. As a consequence, therefore, of the threatening course of development, a new attitude towards the forest has been spreading. Everywhere, in all countries, men with foresight have realized that protection of the existing forests is of vital importance.

The man/forest dualism, running like a red thread throughout mankind's development, has undergone three fundamental changes. Originally, the forest was considered to be man's enemy; later, it became his servant, to be abused, exploited, and little respected; now, each inhabitant of a country, whether forest owner, forest visitor, or one who stays well away from the forest, is gradually being forced to realise that he lives, directly or indirectly, consciously or unconsciously, on the abundant, irreplaceable gifts of the forest.

The forest –
our destiny

It is impossible to separate from one another the forest's contributions to life, for example its useful, protective and welfare functions. The various contributions are not distinctive, nor are they for the greater part measurable or numerically demonstrable. Of many aspects we still know very little. Faced with the complexity, vastness and long-term nature of the interrelated factors in Nature's whole economy our investigations become very complex, time-consuming and to some extent perhaps fruitless. When we come to assess the forest's significance, we should not take present-day conditions alone as our starting point, since these are uncertain enough in themselves. It is much more important to take future developments into consideration. These indicate quite unequivocally a worsening of all environmental dangers and an increase in the forest's worth.

Large forest areas can in many ways influence a region's local climate for the better. The climatic differences between wooded areas and adjacent areas without woodland can be as great as between two mountain slopes facing in opposing directions. A few degrees difference in temperature on a hot summer day is sufficient for a breeze to arise that brings cooler and fresher air from the forest into the hot, dusty city, like the cool sea winds that blow towards the open country from large water surfaces. The wind barrier formed by forests and strips of woodland is also of importance to the local climate. On the windward side of the forest a cushion of air is formed, and in this cushion the wind is considerably less in comparison to the open areas. Behind the forest there lies an area protected from the wind, which up to a distance of ten to twenty times the tree height also provides a perceptible calm. Checking the wind results in the soil drying out less, decreased evaporation in the vegetation, more dew formation and less fine soil blowing away. In many places, for example from an early date in the steppe regions of Russia, wind-protection tree belts were planted transversely to the direction of the prevailing wind.

The forest and avalanches

Snow layers slipping away from steep slopes with elemental force usually cause considerable damage. In the mountain valleys of the European Alps the forest belts in the paths of avalanches have been more or less ripped to pieces. Great avalanches cannot be checked effectively by any technical means. Observation over many years has shown that a mountain forest prevents a blanket of snow breaking away and sliding downwards. It is therefore particularly serious that the former natural timber line was pushed downwards by human interference, so that many new areas of attack were laid open that were formerly wooded. These areas must now be safeguarded by new obstacles if the lost protection is to be restored. Here Nature has shown with startling clarity what occurs if the forest disappears. Great importance is therefore attached to re-afforestation to restore the former timber line.

Forest and water

The close relationship between forest and water has long been realized in all those countries where progressive deforestation has caused high-water danger to increase and intensify. As early as 1863 the biologist and forester E. A. Ross-mässler pointed out in his book "The Forest" that in wooded areas only a small surplus of heavy rain fails to be absorbed, and runs away causing no damage, while the same rain flowing from the naked, rocky slopes of woodless mountains causes dreadful devastation. Because of its loose structure, particularly in the upper layers, forest soil has a greater porosity than open-land soil. Investigations over many years have shown that under European conditions woodland soil is

10–20 times — or as much as 30 times — more porous than open-land surfaces. Forest soil, provided it has not already become saturated following long rainy periods, can therefore absorb, store and slowly discharge again all the water from heavy precipitation and rapid snow-melting, even on steep slopes. In contrast to this, the surface discharge from open cultivated land is great, bringing danger of flood and erosion. This surplus water in no way provides a higher yield of useful water in the drainage area, since the water flows away quickly and is lost to the water supply. Good forestation balances the water economy of rivers. Flow from the heights during heavy rainfall is effectively reduced, whereas low-water flow is increased during the dry period that follows. This abundance that still flows from woodland brooks at low water is of major importance to the supply of water, because ground water is fed by the infiltration of brook water. The effect of the forest is also important when the snow melts. In the shelter of the forest the melting of the snow blanket is delayed, extended over a longer period, and the unfrozen soil allows it to soak in completely. The forest's advantageous effect upon water supplies became even more important as a result of severe injury to the natural water cycle of open land. This injury continues in the reclaiming of large areas of reed and marsh and in the covering of extensive areas with impermeable surfaces (roofing, asphalt and concrete). The manner in which the water supply is quantitatively affected by the forest has not been fully explained. On the one hand, the forest is itself a large consumer of water, but on the other hand the precipitation lost because of direct surface wastage is much less in the forest. E. Kirwald estimates that the storage capability of a well-stocked forest with a root area three feet deep is 5 million–20 million cubic feet per square mile.

The higher incidence of dew and mist in the forest may be of importance. Dew is an important ecological factor, especially in continental dry regions with cool nights. A cautious estimate is that it may amount to as much as 8 ins a year. The Pacific sea mists deposit 6–10 times more precipitation in the pine forests along the coast than in the meadows, and can amount to as much as 40% of the annual precipitation.

Above all, the forest prevents soil erosion. Tree clearance was the first step towards destruction of the soil. Woodland cover not only reduces surface wastage of water, but the dense root system of trees and shrubs and the stable layer of humus also wonderfully protect the soil. Swiss investigations have shown that sparsely-wooded drainage areas produced during a ten-year period three times more soil drift than comparable wooded areas. In river valleys in the United States, erosion following complete deforestation of a large area was eight-and-a-half times greater than in the untouched forest region. It is as indisputable as the advantageous effect upon water economy and upon soil drift that the forest keeps ground water clean. Forest soil is an excellent filtration device, cleaning precipitation water of all its impurities. Because there are no building structures, no seepage of unpurified sewage water, oil or poisons occurs in the forest. Ground water located beneath the forest, and springs arising from the forest, are therefore well protected and healthy. In contrast to this, the dangers inherent in surface water in populated areas and regions of intensive agriculture have increased alarmingly. W. Koch claims that, depending upon the type of cultivation, 10–60% of agricultural soil in the Federal Republic of Germany is treated with chemical agents; in the wine and fruit industry it is 70–90%, but in the forest only 1–2%. Forest is therefore the best insurance for water protection areas. Urgent representations must however be made to the forestry industry that it should stop using pesticides in the forest. From the viewpoint of water economy, the most important duties of the forestry industry are to preserve or increase the water yield of certain drainage areas, to preserve water as a commodity, to promote proper water control and to protect the soil against erosion by water.

The struggle to keep water healthy has recently also involved large international organisations, such as the World Health Organisation (WHO), the Organisation for Economic Co-operation and Development (OECD), the European Economic Community (EEC), the Food and Agriculture Organisation (FAO) and numerous other international bodies. The Council of Europe has summarized the most important results in its "Water Charter". This fundamental declaration records that plant cover, and especially woodland, plays an essential part in the preservation of water supplies.

Forest and air

Pollution of the atmosphere by smoke, soot and dust has in many densely populated areas reached such proportions that it causes not only inconvenience but a health risk to the population. Each of us breathes in every day 420 cu ft of air, and in this way in a large city (with up to 3,000,000 dust particles per cubic inch of air) we take in ·0007 of an ounce of dust. The larger particles are kept back by the upper respiratory passages, but the smaller dust particles reach as far as the lungs, and to some extent the blood. Forests and well-graduated tree belts make an excellent contribution to the cleaning of the atmosphere. In addition, not only large forests but also smaller woods check wind-speed, so that the coarser dust becomes sedimented. Wind masses sweeping across large forests drift downwards more rapidly over the forest because of the lower temperature. The dust in the air flowing through the forest is trapped by the tree crown: Laboratory investigations have shown that one acre of pine forest possesses a capacity to filter up to 13 tons, and one acre of beech forest up to 30 tons.

148

Rain cleans the filter once again; the filtration provided by deciduous trees is always completely renewed after the annual fall of leaves. Radio-active substances are also very effectively filtered-out by the forest. It has been shown that the foliage of trees on the side of the forest away from the wind displays radio-activity as much as four times less than on the exposed side. Crops in the shelter of strips of forest display only one-fifth of the radio-activity of plants on the exposed side, in some cases as little as one-twentieth. It is, however, more difficult to remove air-borne contamination. The forest is itself one of the most sensitive forms of vegetation. Even though leaves and needles of forest trees far away from sources of emission may display a high toxicity, the forest is unable to contribute significantly to air decontamination. But it does present an obstacle to flow, causing increased turbulence with consequent dilution of the toxic agents. Hedges and wide belts of forest provide residential areas and agricultural land with significant shelter from exhaust gases.

The forest's contribution to air regeneration by the decomposition of carbon dioxide and replacement of the oxygen used in breathing and other forms of combustion is very important. Many people are now saying, however, that the oxygen supplies present in the air are inexhaustible. But if we remember the progressive increase in the consumption of fossil fuels and the vast expenditure of effort, under the most difficult conditions, to find new sources of natural gas and oil, it may be foreseen that the combustion of fossil fuels, and thus the wastage of oxygen supplies, will develop exponentially. The exploitation of fossil materials is being carried on at a rate one million times faster than their original natural formation. It is a fact that the carbon dioxide concentration in the atmosphere in densely populated areas has measurably increased, showing that the vegetation cover can no longer process the carbon dioxide produced. Motor vehicles in the United States burn twice as much oxygen as the vegetation cover of the North American Continent can produce. The Special Committee set up by President Kennedy to investigate problems of environmental pollution predicted that the original carbon dioxide content of the air would have increased by 25% by the year 2,000 if the consumption of fossil fuels continued to increase at the current rate. This would constitute a geophysical experiment on a gigantic scale whose consequences, climatically and biologically, are inconceivable.

Up to 70% of the oxygen in the air is produced by sea plankton. But the oceans' oxygen production has fallen by one quarter in recent years because of increasing pollution. For the protection of the natural oxygen/carbon dioxide cycle, forests and open vegetation areas therefore become ever more important.

Protection against noise

The stillness of the forest is especially refreshing to people suffering from stress caused by noise. Man needs quietness to relax. Extensive woodland is therefore of major importance, since only this offers true stillness. Protection of residential areas against noise is also becoming increasingly important, but it is not sufficient to plant a few trees in these areas. On the other hand, even a few rows of trees have a measurable noise-damping effect. However, a satisfactory result can only be provided by belts of woodland about 100 yards wide, which bring down the maximum noise level to the average level of areas with little traffic; the trees have their greatest effect on the high-frequency ranges, which are particularly annoying.

Planning of settlement areas.

Traditional settlement patterns in the countryside adhered to their original characteristics. In contrast to this, short-term building expansion in the most suitable settlement areas led to the menacing problems of over-population. The stone deserts expanded unrestrictedly in all directions and smothered the green landscape. In Switzerland, only the forest, its well-being protected by law, was able, with its islands of greenery, to hold back the stony tide. Only in very few countries did this occur. Elsewhere, wherever the forest stood in the way of settlement it was levelled.

Forests near to settlements represent the most effective and the cheapest green areas and buffer zones. Unaesthetic and unattractive tall buildings, monotonous and lacking imagination, have made the cities uninhabitable through the noise and disturbance which they engender. Here are born those problems that face us ever more clearly. This growing estrangement from Nature causes more and more people to move away from the city, to some extent taking their problems with them to the previously peaceful open countryside. At a time when man ignores Nature's frontiers, the forests and waterways form the only spatial elements that somehow still have permanence. In Swiss land-planning concepts, the forest, as an independent zone, is employed as a firm framework for area planning. The more uncontrollable is the progress of technology and motorization, the larger is the space required for settlements, industrial areas and road construction. The more the landscape is turned to stone, the more important becomes the remaining green space near to cities. Of these spaces, the forests are, without any doubt at all, the most important.

Forest and landscape

The forests shape the landscape and imprint their image on it in a characteristic way. Woodland is the most impressive form of vegetation. It is not only large forests that are of importance to the landscape; copses in fields, hedges and stands of trees along the banks of streams are important elements in the total picture, and have therefore a significance far beyond their share of the landscape's area.

The old cultivated landscapes developed harmoniously, although frequently on a foundation of early beginnings that happened to be attractive. Our forefathers had a sense of proportion and of blending one thing with another. The Industrial Revolution led, above all in the twentieth century, to harsh, almost chaotic, transformation of the former landscape. High-rise building developments, extensive industrial complexes, great technical installations and exploitation of the waterways have chiselled strange features into the face of familiar landscapes. Values were sacrificed all too easily to the demands of traffic and house building, usually involving the loss of all natural beauty in the landscape. Technology can be cruel to Nature and countryside. Forests large and small, as well as copses and groups of trees, are often able mercifully to conceal these disfigurations. Forests reduce the countryside's vulnerability. The delightful alternation of open country and forest is of immense value to the beauty of a country. Large forests surrounding a residential area symbolize most impressively the feeling of security, homeliness and togetherness; the many little woods, for their part, often contrive still to give to landscapes affected by technology the character of a countryside close to nature. The forest is therefore indispensable to the protection of the appearance of the landscape. Vast forest areas are immensely impressive, where into the far distance ridge after ridge of woodland

rise like ramparts, becoming progressively bluer until they fade away into the hazy distance, a last soft mistiness marking the horizon. "As far as the eye could reach there was to be seen no other picture but the dissolving forest, spread over hill and dale, extending to the remotest point of vision, leading far away to the sky, radiant and blue like its sisters, the clouds" (Adalbert Stifter).

Natural cycles

Nature has its closed circuits, marked by a stable equilibrium. There exists between forest soil, air space and tree growth a cycle of unchanging stability, and all substances participating in this cycle may pass through it, often without restriction. The forests are not commercial undertakings that need artificial energy, consuming raw materials and producing useless waste. They produce a precious substance — organic wood. Wood is the only raw material that is uninterruptedly reconstituted by photosynthesis. Wood is also the only material whose production needs no artificial energy and no fertilizers, and which at the same time regenerates the atmosphere. The problem of waste, which increasingly troubles modern industrial society, is solved in the forest in the most wonderful way. All organic residues are decomposed by natural forces to form their original substances, and reintroduced into the cycle. Nothing is wasted. Nowhere is the indestructability of matter and the cycle of elementary substances more evident than in the noble process of the growth of the forest.

Thus had Nature worked tirelessly for millions of years to create a stable, steady system, until man, in the industrial age, recklessly interfered. He produced great quantities of substances alien to Nature, and resistant materials, persistent poisons and radio-active waste that cannot be decomposed, that can no longer be integrated into the production process and that strain the whole economy and endanger the environment.

The water cycle is imposing in its dynamism. A large proportion of precipitation is retained in the forest in the crowns of the trees, penetrates into the soil or is evaporated, so enriching air humidity. The rest flows via brooks and rivers back into the sea, its original source. The humidity of the air again forms precipitation. This interchange continues incessantly, and at the same time keeps in motion those functions associated with it. It is estimated that oceans, fresh water lakes and the mighty areas of vegetation evaporate about 500 billion tons of water a year, an average of about 16 million tons each second. The evaporation of these enormous quantities of water, which fall back to earth as precipitation, absorbs about 10 billion calories of solar heat a second, equivalent to the thermal value of 1.6 million tons of coal. The surfaces of oceans, lakes and rivers are greater than the land areas of the earth. The abundant plant cover can, however, return more water to the atmosphere than a water surface of the same size. Water is supplied to the air by the crown area of forests more continuously than from meadows and fields. The role of forests, "these bridges between heaven and earth", in the earth's water cycle is therefore especially crucial (*Implosion*, Vol. 4, 1962). The steady regularity of the evaporation process has been seriously disturbed in the vast deforested regions.

During the growing period, the green crown uninterruptedly produces sugar and starch from carbon dioxide and water, and at the same time life's vital oxygen. Wood, branches and foliage fall into decay and are broken down by bacteria. This process consumes oxygen and produces carbon dioxide; it forms the reverse process to photosynthesis. The roots extract inorganic nutrients from the soil for the tree's growth. In spite of thousands of years of production, the soil does not become impoverished. The mineral substances are continually replaced by the disintegration of soil components and the decomposition of organic waste. The forest's nutrition cycle takes place with amazing rapidity. In deciduous forests, almost a complete cycle is completed within two years; in tropical forests the period is even shorter; in coniferous forests it takes 3–4 years. In the jungle, the cycle is without beginning or end. In commercial woodland, part of the nutrients used are irreplaceably extracted in the timber harvest. Trunk wood contains only 0.2–0.4% of mineral substances and 0.2% of nitrogen, but needles and leaves have 1–3% and 3–9% of mineral substances and 1% and 1–3% of nitrogen respectively; these are left on the ground and are decomposed again, so the loss is very modest and is abundantly replenished in the active soil by decay. Forest soil should therefore never be fertilized and neither should it be cultivated. Nature takes wonderful care of nutriment replacement and soil loosening, and needs no assistance.

Conservation of plant and animal life

It is true that most forms of life have the physiological ability to continue to thrive over vast climatic and vegetation areas. But the competitive conditions present in any community restrict the incidence of most flora and fauna to those places that especially appeal to them. If these habitats are destroyed many plants will disappear, but for animals there is more possibility of retreating to an adjacent and similar area.

J. Dorst has said that progressive deforestation is the most apparent indication of large-area biotic changes. But the expansion of residential areas and of transport networks has also irrevocably destroyed many unwooded biotic areas that form the natural environment of flora and fauna. Industrialized agriculture also ensured that large biotic areas were lost. In all cultivated areas large marshy meadows were drained by irrigation, and pools and bogs were filled in. Not only were these elements of the landscape important to water economy, not only did they possess a peculiar beauty, but they provided a specific habitat for the most varied, and ecologically valuable, communities.

Investigation has shown that at least 120 mammal species and 150 bird species have, through man's rapacity, been evicted, never to return. At the preparatory conference to plan an international convention for the protection of threatened flora and fauna from extermination, it was stated that at present about 800 mammal species are indirectly or directly threatened with extinction. The fate of many animals is already decided. Nature is on the retreat everywhere

But it is true that in all eras flora and fauna have become extinct as a result of natural changes without man having been able to prevent it. On the other hand, in the long term new species and genera have appeared. These gradual changes, however, have nothing in common with the impoverishment of natural fauna and flora caused arbitrarily by man, and completed within such short time-spans.

Since forests large and small are among the communities most rich in species, they are of paramount importance for the conservation of flora and fauna.

The balance of Nature

Forests possess a great capacity for regeneration when they become the victims of disturbance. When, for example, a species of insect multiplies excessively and causes damage, its enemies will also then multiply to ensure that no imbalance occurs. The richer in species is the community and the more interconnected are the cycles, the greater becomes the capacity to regenerate and the stability of the ecological system. "This richness in species of animal and plant, the number of ecological niches and their entwining, continually in-

164

creased in the course of evolution until Industrial Man – unaware and unintentionally – interfered by employing processes that frequently destroyed the complex framework of creation'' (E. Basler). Landscapes made up of many biotopes are more stable in their balance than biologically uniform landscapes. Unfortunately there are still serious gaps in our knowledge of the complex, scarcely explored total economy of Nature. With increasing specialization, it is all too easy to neglect the often unforeseeable and mostly unknown side effects, frequently serious in their consequences, that are necessarily associated with every economic activity. More thorough ecological research and more profound ecological thought is therefore essential. The preservation of the environment should not be left entirely to politicians and technologists. The forest's special properties – its rational nutriment economy, controlled water cycle, promotion of microbial soil processes, regeneration of the atmosphere and richness of fauna and flora – make an indispensable contribution to the ecological resistance of the countryside. Forests evolved by Nature can themselves take care of disease and injury. After a disturbance of any kind a state of equilibrium is restored with the passage of time, even though Nature may be forced, when necessary, to make a detour via different stages of regeneration. Each species in the community helps to preserve the ancient law of Nature – balance in the community. The forest is therefore one of the most stable elements, biologically and ecologically, of the countryside. This becomes even more important because open country's capacity to regenerate has suffered severely at the hands of monoculture, artificial fertilizers and chemical pesticides.

The forestry industry has an obligation not only to ensure the conservation of forest areas but also to cultivate in a manner that accords with Nature. It must refrain from all interference that goes against Nature, leads to injury to the soil, weakens the structure of a stand of trees, destroys a species of flora or fauna, or impairs the natural cycles. Complete deforestation, or even the introduction into mixed forest areas of tree species alien to the locality, represent a flagrant disregard of ecological continuity. Even in those cases where, after complete deforestation, the area is immediately planted again, it takes many decades to re-knit the torn threads. In most cases it is probably impossible for the destroyed forest community's original abundance ever to be restored. It is therefore ecological folly to replace the rich mixture of natural deciduous trees by a few species, however commercially valuable they may be.

The beauty
of the forest

The Age of Enlightenment in the 18th century brought with it an intensive analysis of Nature. It led to the discovery of the Alps and finally to the discovery of the forest. Under the influence of eminent research scientists and Nature-lovers, the public began to take an increasing interest in the forest and to realize its possibilities.

The vast green area of the forest with its enormous columns of trees is like a church interior that instils into us a reverence in which nothing disturbs the thoughts that pass through the mind. Stillness, growth and permanence prevail here. The development of the forest is symbolic of the rise and fall of great empires; the growth of the tree symbolizes human life. But only he can have true empathy with the forest who carries within him the inner will to do so, to absorb this beauty and magnificence. In our attitude towards the forest, or towards Nature generally, it is necessary that we should have an inner ethical harmony, a respect for life, a respect for creation and for the Creator, and finally a religion. The more we surrender ourselves in Nature, the more we sense the complex, indissoluble ties which we shall probably never be able to penetrate to their full extent, and of which we also form part. In our intercourse with the forest, we must learn to look, observe, think and enjoy silently. It is not so much the individual shapes or sounds that we perceive; it is the entirety of the infinitely varied impressions, the interplay of all we see, hear and sense that impresses us. As well as all this there is the exhilarating feeling of security, of being part of the community. We are challenged to stop and ponder on those of our actions that are in one way or another questionable. Especially today, when we are running the ever-increasing risk of alienating ourselves from ethics and ideals, such hours of self-reflection are more necessary than ever before. In such silent

meditation we question the meaning of life; "that I may see what holds the world together at its very soul".

The observer whose attitude towards the forest is one of great love for all that lives will enjoy innumerable wonders. Here there moves, crowds and lives a thousand-fold life in a thousand-fold forms. The forest is one of Nature's most mysterious creations. To be one with the forest there is no need to know all the plants and countless animals and insects, although we may be encouraged to learn more about the many things we see. If we check our steps, our feet coming to a halt in the rustling foliage, we shall hear that the forest is never quite still. Somewhere over there something murmurs, perhaps a little brook; there is a quiet humming in the air; a leaf rustles; in the mossy carpet or among the weeds something is busily crawling; somewhere a woodpecker hammers away; a jay gives its warning cry; a dead branch falls; a quiet breeze whispers through the tree-tops. All this makes up the voice of the forest; a varied, gentle harmony. Only rarely is the stillness penetrated by a weak sound from the outer world, whose faraway roar only enhances the sense of quiet and solitude. A little time spent in the forest, whatever the season or time of day, presents people seeking relaxation with an experience of unequalled richness. Compared with technology and commerce, Nature offers us values that are as important to our well-being as material goods, but whose final disappearance is threatened. It is therefore one of the most honourable duties of parents, teachers and older people to arouse in the young an appreciation of the magnificent, the unique and the wonderful that is inherent in every tiny creature, in a beautiful flower, in a tree, in a stretch of countryside, in a forest; a thing that we cannot ourselves create but can only destroy.

The forest in art and literature

Artists have long tried to express the mood of the forest in words, colours or sounds. Poets, painters and musicians have always been attracted towards the forest, finding inspiration in its stillness and solitude.

Countless poems have the forest as a theme. The emotional relationship between man and forest was first expressed in the old fairy-tales. In these, the forest is still what it was to our forefathers, when little settlements lay like tiny islands in the vast forest. We still encounter in these fairy-tales the world of mystery. Legends and fairy-tales are some of the most precious ingredients of our cultural heritage. In later poetry the forest is represented more as a grove, used as a backcloth to the play of allegorical phantasmagoria. In the medieval view of Nature, the forest was a secretive place where lovers met; but it was also the refuge of the transgressor, the penitent and the hermit. Eichendorff especially represents the forest as creation's imperishable perfection. But many other poets until the present day have described forest scenery and vistas, and all the emotions and thoughts that moved them. But no poet has described the forest, the woodland, the forest people, and the nostalgia of forest and homeland with more devoted and loving sensitivity than Adalbert Stifter.

The veneration of tree and forest has also found much expression in the pictorial arts. We know of stone reliefs from earliest times that include representations of trees. The old Gothic cathedrals symbolize the cathedral-like interior of the tall forest. Plants and trees were repeatedly used as motifs at the peak period of Gothic art. Trees are often to be found as the tree of life in pictures dating from the early Middle Ages. The forest also appears in many early landscapes, but we are more impressed by the trees and forests painted by the masters of the early 16th century, the Classical period and the Romantic period.

The forest in all its forms has also been intensely reflected in music. The forest theme is continually repeated in the minstrels' songs, in many folk songs, in the songs and music, and later in the operas, of the great composers. Surely the forest is itself music! The song of the birds, the murmur of the brook, the rustling wind and the gale that storms through the tree tops make up the forest's *leitmotiv* of song and sound. The inspiration of many composers had its birth and was matured in the sonorous solitude of Nature and the forest.

The rhythm of the seasons

The endless, eternal change of the passing days and years is full of fascination, and its impressiveness grips us anew each time we experience it. In temperate zones there is a clear climatic rhythm in the seasonal events. The deciduous forest with its manifold colour tones has infinitely richer seasonal variations than the austere, dark coniferous forest. The rhythm of the year is more clearly expressed in its changing foliage. The coniferous forest is sterner, more solemn, almost gloomy in overcast weather. Only in spring, when deciduous trees are already decked in brightest green, does the friendly, fresh green of the young shoots cover the dark mantle of coniferous trees. Only the larch insists, after the bareness of winter, in clothing itself in the brightest spring-green.

After the cold and starkness of winter, Nature's awakening in springtime is indisputably one of the glories of the yearly cycle in our latitudes. Who is it that brings the message of spring down to the darkness of the earth, where in the first spring days life is already awake? Who tells the countless creatures that spend the winter in a sheltered hiding place to come out to enjoy the warmth of spring? Spring in the deciduous forest, where so many secrets become plain for all to see in the forest community's vast interplay, is of unsurpassed beauty.

In the light of the early spring forest, an exuberant covering of spring-flowering plants shoots upwards in the first warming rays. It is a profuse growth, a busy germinating and flowering. But their time is short. With the coming into leaf of the trees, the zenith of their life is past and the plant cover merely leads a life in the shade. This is why the spring picture is the most impressive and joyful. The forest changes from day to day, becoming more beautiful, more colourful, richer in life. The animals also awaken, and their life becomes busier and noisier than at any other time of the year. The adaptation of the flowering plants in the deciduous forest to these few sunlit weeks is a wonderful thing. During the previous year, they stored reserves in bulb, rhizome and root to make it possible for them to participate in this race against time. Their seeds ripen even before the advent of early summer. Their underground storage organs are filled again for another spring. Only the development of vegetation from below and upwards at different levels safeguards the right to live of all members of the forest community.

As summer approaches it becomes quieter in the forest. This is the time of development, of growth, of production, of the formation of fruit and seed, wood and storage materials. The birds become quiet, and only the insects fill the dusky spaces with their humming. The forest stands peacefully in the heat of the day. The cool, shady darkness is pierced by only a few rays of the sun; the light has an infinite variation of tone. In the high summer stillness we hear only a gentle humming, a slight movement in the leaves and a coming and going on the forest floor.

Autumn presents the second climax in the life of the forest. The days become strangely transparent, the colours more gentle, the light more diffused. This is the long period of preparation for the coming farewell. Nature is liberal with its last surge of the joy of living. The crowns of deciduous trees and the shrubs are resplendent in their colouring, from burning red and radiant yellow to warm brown, a magnificent contrast to the dark crowns of fir and pine. The colours, like musical tones, swell into a harmony of unsurpassable beauty, fading quietly away in the death of falling leaves.

The sun is lower on the horizon, and we sense more and more that Nature is about to go to its rest. The trees' activity is decreasing, the cycle stagnates, growth comes to a halt. At the base of the leaves the supply lines are sealed by a thin cork coating. Soon the leaves will begin to fall. The plants of the herbaceous layer, so green in summer, are turning yellow; life draws back into the underground roots. Only the coniferous forest stays green throughout the winter. Here, only the oldest needles die away; their passage to the ground is so quiet that it can scarcely be heard. Also scarcely perceptible, the animals' flight to their winter quarters is now in progress. How do they know it is time to seek shelter against the severity of winter?

The forest in winter also has its own peculiar charm. Life in the forest has come to a halt. Snow flakes come to rest silently on trees, plants and soil. The stillness is increased by the mood of sleep and rest. But long ago, even before the leaves had fallen, were formed the buds from which the fragile leaves will break out again. Reserves were stored in the trunks for rapid liberation in the coming spring. Winter is a hard time for many animals. They lack nourishment then, and many of them sleep through the cold season. In the warmth of the soil, in the soft moss, under the bark and in the cracks, in hollows and holes, in old tree trunks and under dry leaves, everywhere are resting the animals whose vitality is reduced to extreme torpor. Many, thus protected, survive; but many must suffer bitter hunger and die or become the prey of other animals. And yet, after the long winter months, all Nature will awaken again from its deep slumber into a new life. Nature knows many ways of adapting itself to this alien season.

181

There are families of insects who reduce the body's water content during winter so that they can survive. In especially cold climates some form additional glycerin. Winter can then be as hard as it wishes; survival has been ensured.

The rhythm of the time of day

No joy can rival the awakening of the forest in the early morning of a beautiful day. At first, it is still night under the trees and peace reigns in the forest. Gradually a weak band of light burgeons in the East. In the forest darkness retreats with hesitating step, and as yet there is no forewarning of the brightness of the coming day. Then the first cry of a bird sounds; more and more awake and join him, until a many-voiced song rings through the vastness. Everywhere in the forest, on the soil and on the branches, the most diverse of all activities is coming to life. Ever more sounds and voices join the chorus, multiply, increase and unite in the forest symphony. At last the rising sun slants through a few gaps in the forest roof. The new day has arrived. Everywhere there is a busy hurrying and scampering and humming. Invisibly and inaudibly the trees are at work, water and nutrients rise in the trunks, evaporation and assimilation is going on in the green crowns, millions of cells are being formed, branches and roots stretch outwards, wood cells store themselves in the trunks, flowers give off their scent, and gently the wind rustles through the branches. So it goes on all through the day, until the light of the evening sun glides across the branches and dusk falls quietly on the interior of the forest. Secluded stillness again prevails, the forest stands reverently like a silent prayer, while from the edge of the forest the evening fire of the sky glows into the interior and slowly fades away. Then comes the all-enveloping night.

182

It is a unique experience to walk through the forest in the darkness of the night, an endless black wall before us, a single star twinkling from inconceivable infinity through a gap in the crown. Even in the depth of the night the forest is never completely still; there are hushed sounds when the night population sets out foraging for food or prey. Strange noises can be heard among the whisper of the forest night: a tapping in the dark, the quiet cry of a bird, the discreet noise of a stealthy creature, a sudden rustle, the merry flow of a never-resting brook. Only hesitantly do we continue on our way, step by step. In this magical mood we sense with reverence the song of the forest, the voice of Nature.

The rhythm of the weather

The changes of mood in the forest are accentuated by the whims of the weather. The forest is beautiful not only in fine weather when all is song and rejoicing and the bustle and hum of happy activity, the sun's rays falling upon the green carpet or upon flowers still in bloom; it also has its beauty on a grey, rainy day when the green, dripping space is wrapped in dusky light. Steady, monotonous rain increases the feeling of solitude. Everything is damp. Water trickles from the trunks, the wet leaves glisten, the soil gives off the heavy, humid smell of fertility.

A storm in the forest can be overwhelming, thrilling, almost frightening in its power. First there is a gloomy twilight in the uncanny stillness. The trees stand silently awaiting the approaching combat. The leaves scarcely move. Beetles disappear, creatures seek shelter, birds huddle in their nests as it grows darker and darker. Distant thunder murmurs, lightning cuts through the black clouds. Then suddenly the forerunners of the approaching thunderstorm drive through the tree crowns. Wave upon wave bends the treetops to and fro. It blusters in fierce attacks, increases to an endless roar and then rushes over the forest in sinister turmoil like the surf of a raging sea. A surging and roaring, a moaning and groaning, fills the vastness, while the animal world, in dumb fear, listens to Nature's wrath. The noise of the lightning and thunder is as if the earth is soon to be torn asunder. Motionless, we watch the forest's resistance to the force of the storm. We sense the immensity of the root network with which the trees claw their way into the soil to avoid disaster during the threshing and the surging. It becomes dark as night, while the rain beats downwards. After a time the rage of the storm subsides. It moves away, ending in a quiet, soothing rustle. Distant thunder still echoes from afar. Slowly the rain stops; but countless drops are still shaken down from the treetops by the last gusts of wind; leaves, trunks and grass glisten; the forest soil steams. The anger has passed; the forest is quiet once again; the evening sun is aglow, casting its golden rays through treetops wet with rain; the first tiny creatures slowly venture out again. We leave the forest in the evening light, wet through, still numb from this mighty performance and sunk in thought.

He who understands the language of the forest cannot avoid being profoundly moved. Forests and trees tell him mysterious legends from far-off times, reveal to him a glimpse into impenetrable secrets, shower him with beautiful dreams and in their steadfastness point towards a far distant future. Then we can reason away our sorrows; they lose their importance if we see our troubles in relation to this infinity and magnitude.

Living with the forest

Hundreds of millions of years were needed by Nature to evolve the living creatures inhabiting the earth today. But it took only a few centuries — often only a few decades — for man, fashioning Nature to his will in field and forest, irrevocably to destroy much of this. Every forest, every bog, every marshy meadow, every stretch of countryside sacrificed by man to his machines was a destruction of part of his habitat, of his soul. The demands made on the forest are greater today than ever before. We expect the forest to supply us with a raw material that Nature always produces anew, but which we exploit to an extent that Nature is unable to match in its production. At the same time, the forest is expected to maintain its ever more important protective and social roles. The forest is expected to provide space for expanding settlement in many population centres and, simultaneously, for the extension of farming in primitive agricultural countries. Even in the European countries with their well-developed forestry services, new intrusions into existing woodland repeatedly occur. Preserving the landscape unchanged in every respect need not be the main consideration. But at a time when everything the world over has been shaken by severe upheavals, when the conflict between development and growth on the one hand and conservation and preservation on the other rends people apart, the quiet continuity of the process of Nature represents the only stability to which we can cling.

The classical protective role of the mountain forest cannot be replaced by any technological means. The Alpine region will become more and more a recreational area for the inhabitants of over-populated centres. This trend will lead to more settlements in the mountain valleys, more building in the holiday resorts and denser development. As a consequence of this, the prevention of avalanches, landslides and rock falls and the restriction of loss of precipitation into the valleys becomes so crucial that not only will existing woodland have to be preserved and cared for but additional protective woodland will have to be provided in areas at risk. The forest's protective function will increase even more in importance, since protection by woodland is essential if the exploitation of mountain regions for recreational purposes is to be intensified.

In the over-populated areas, the forest's recreational role will be even more vital. Civilized man's life has become unnatural; it is allied to a growing estrangement from countryside and Nature, and it endangers man's health. The demand for regular relaxation in recreational areas close to Nature and free from fumes has reached amazing proportions. The forest offers for recreational purposes the optimum physical, climatic and hygienic conditions. When the visitor from the noisy city, the worker escaping the strains of the rat-race, enters the eternal beauty of the forest, where soothing green prevails and the vast, silent space surrounds him, he has a deep sense of inner relief that he has escaped the physical and psychological stress. We feel beneath our feet a floor so different from the hard pavement; we breathe an air wonderfully clean and fresh; we smell the indescribable scent of flowers, berries and mushrooms, the aromatic odour of resins and tannin, the mouldering smell of the damp soil, everything that makes up the forest's peculiar perfume. Gradually a marvellous abundance of shapes and colours is revealed to the wondering eye. Consciously or perhaps unconsciously the ear catches a plethora of soft sounds, which together form the soothing chords of the forest song. The whole fullness of life, to which only the forest has the secret, becomes clear to us. Many people certainly avoid the forest; they are in fear of the stillness, of being alone, since from this might emerge a moment of reflection, a sense of inner emptiness and of one's own hopelessness. But for most visitors the forest becomes an adventure, a flight from the restless world into the continuity and harmony of Nature. Here, everything that happens depends only upon the sun's position in the sky. The clock, obligations and appointments have no place. Endless is the stillness of the forest, a stillness without anxiety, a stillness that is not silence because the forest lives, breathes and works to the rhythm of Nature's cycle. When we begin to comprehend Nature's dominion, to sense the interdependence of all life processes, it is then that beautiful and reverent thoughts enter our hearts, and we yearn for something that we ourselves cannot identify. We feel a sense of devotion towards primeval Nature, whose creation we are. Adalbert Stifter declares that in late summer "the best the forest gives us is that we are alone, completely alone, surrendering ourselves completely".

Re-stocking

All over the world there are deforested and destroyed areas of inconceivable extent. The FAO is trying to initiate regeneration in the deforested countries. It will be an important and difficult task, in the developing countries especially, where the overall economy must be built up, to transform wholly deforested areas into forest once again in order to arrest and eventually to repair the immense damage that now grows with horrifying speed. Whether we shall be able to reconstruct the wreckage of the past, or shall continue to destroy what has remained, will determine mankind's life tomorrow.

Very many countries have already re-stocked large areas. But this is a difficult enterprise, and it calls for immense patience, large sums of money and a long time-span to redevelop a forest on the weakened, barren soil, often worn down to the rock foundation. Even if planting is successful, summer drought, hard winter frost, soil blown or washed away can still lead to failure. Often the local tree species suitable for re-stocking are only sparsely represented in the regions concerned. Israel, for example, has tried out many tree species from other arid

196

areas of the world with similar ecological conditions to examine their suitability for restocking the deserts. Most of the first plantations will show poor development. Two or more tree generations are required to procure once again the fertile soil on which good forests can thrive. Nowhere more than here is it proved that it is quicker to destroy than to reconstruct. But nowhere more than here can it be demonstrated that man has no other recourse but to heal Nature's wounds, which he himself so thoughtlessly inflicted. It is particularly difficult to recreate the forest at the frontier of its natural habitat, for example on the edge of the desert or at the upper timber line.

The Russians were the first to plant wind protection belts near towns and villages. In the period before the First World War, 270 square miles annually were restocked; since the Second World War this has risen to 1,150 square miles. Under the General Plan of October 1948, 23,000 square miles of new woodland will be developed in the arid regions of the country. In the form of protective belts, these areas will cut across the entire European Russian steppe region; some belts will be as long as 600 miles. On steep slopes, forest belts will be planted transversely to the fall line to prevent washing away of the soil. While on the one hand these gigantic efforts were in progress, deforestation is on the other hand continuing in the densely populated and industrialized areas.

Over the last 30 years, Spain has re-stocked 7,750 square miles of barren land. About a quarter of this area had to be restocked a second time, since the unfavourable climate did not allow the first planting to develop. The devastating forest fires, which destroyed 750 square miles of woodland between 1962 and 1967, are a great anxiety. Under the most difficult conditions imaginable, Yugoslavia created 40 square miles of new forest in the Karst after the First World War, and by the end of the 1940's a further 950 square miles. In some places, furrows 15 ins deep had to be cut into bare limestone and filled with soil to provide the young plants with the means of growth. In W. Germany nearly 7,750 square miles, or a quarter of the woodland, is no more than inferior scrub. Since the development of integrated forests was made impossible by grazing, the government has ordered, in spite of strong opposition, a temporary prohibition of goat-grazing. Large areas of scrub were levelled and fenced. From the tree stumps are growing slim, budding forests in which the planted black firs are growing superbly. With the immigration of the Jews in the second half of the last century, Israel began to restock large areas of stony desert that had in the past been woodland. Planting became particularly intensive after the foundation of the State of Israel. Up to 1948, 15 square miles of new woodland was developed, but since then 200 square miles has been added to this. In Northern Europe, Denmark, poor in forest, has almost doubled its woodland in recent years. Sweden and Norway, both more than half covered in forest, are planning new afforestation — Sweden 1,500 square miles and Norway 3,000 square miles. The United States founded its Soil Preservation Service in 1953. This service, with its gigantic budget, is trying, by the re-afforestation of great protective belts, to arrest erosion of the soil by water and wind. A British-French group has a project for the Sahara desert envisaging the planting of protective woodland 3,750 miles long and 30 miles wide along the desert's Western edge to check its further advance. Even in China's vast steppes there is an attempt to re-stock some areas. In the last 25 years, New Zealand has planted new woodland over 500 square miles in extent.

Even while these difficult attempts at re-afforestation are being made, in an endeavour to heal the thoughtlessly-inflicted wounds of the past, the plundering continues in most continents. "Blindly we race ahead without regard for the red light. We rush towards the abyss when we continue to rape the earth" (R. St Barbe Baker, 1957).

The forest must be preserved

In the present time of misery and danger, conservation of the natural foundations of life — soil, water, air and Nature — has become the most urgent and crucial task which now faces us. In spite of the attempts in many circles to gloss this over, it is indisputable that the danger to the environment has assumed proportions that force us to take action.

In the future, the forest will be a crucial factor in environmental protection, since it offers the best possibilities for overcoming the problems that threaten us. The importance of the forest to agriculture, to soil conservation, to regional planning, to climate, water and air and to flora and fauna has increased at an unbelievable rate in recent years; it will increase still more in the future. The time is therefore no longer distant when on ecological, hygienic and sociological grounds the forest must be totally protected.

Woodland destroyed in the past cannot be restored on the same scale. It is therefore all the more important to preserve what remains. Man, who destroyed the forest over centuries, who pushed it back wherever it was too near to expanding settlements or was otherwise in his way, who exploited the forest and levelled it by fire, must today protect the remaining woodland by all means in his power. But the forest is not only an indispensable foundation of human existence; in conserving it, the essential habitats of flora and fauna are also at stake. In Nature's economy, in one form or another, in one location or another and at one time or another, these life forms must fulfil tasks that are not to be underestimated. Conservation of the forest will be especially important in the over-populated areas. Here therefore the bitterest struggles and the strongest arguments will become necessary to ensure that no further woodland is sacrificed to residential, industrial or transport purposes. Such sacrifices are not made good by restocking remote areas already extensively covered in woodland, where no new forests are needed and where, on the contrary, indispensable open recreational areas would be sacrificed.

In Switzerland, woodland is protected by the Federal Forestry Act of 1902. It is surprising that as early as the seventh century the legislature issued a ban on forest-clearance. Flood catastrophe may well have threatened large areas of the country, but the danger to the natural environment on its present severe scale could in no way have been foreseen. "Only the fact that our forests are protected by such a harsh law, basically interfering with the Cantons' authority, permits us today to make a fairly optimistic forecast for conservation of the environment in Switzerland" (K. Bächtold, 1971). It is certain — and the efforts in neighbouring countries show this clearly — that today a legal provision of this kind for conservation of the forest could scarcely be introduced without compensation being paid to those affected.

To defend the forest in an insecure world necessarily means to argue incessantly and unyieldingly, in the face of the hard facts of economic life and of conflicting interests, and with very little room for manoeuvre. But the more that mankind's environment is threatened the more urgent it becomes to keep intact the unspoilt countryside we still have. Nearly 2,000 years ago Pliny came to the conclusion that the trees and the forest were the most sublime gift with which Nature had endowed man. Must not this perception be much more compelling today? Forest and man; both are dependent one on the other for their common well-being. "Husbanded forest and man — neither of the two can exist without the other in the face of mankind's policies of cultural and population development" (J. Graf and J. Weber, 1965).

The association of "Men of the Forest" passed the following unanimous resolution in London in 1956: "The famine that still exists all over the world reveals man's dependence upon the tree. The disappearance of woodland and the encroachment of desert upon sources of nutrition advance 30 miles a year along the 375-mile fronts of three continents. In the face of these facts, the men here assembled from 24 nations ask the UNO to participate in drafting an international forest convention."

The Declaration of the Environment Conference at Stockholm in June 1972 came to the following conclusion: "We have arrived at a point in history where more careful consideration must be given to our actions all over the world. The conservation and improvement of the environment for the present generation and future generations has become an objective for all mankind." One of the most important recommendations of the action plan is the conservation of global forest stands. The well-being of mankind is vastly dependent upon the existence or lack of woodland. There is consequently no alternative for us but to live with the forest, to care for it and to conserve it. In a civilization dominated by technology there is nothing greater, more beautiful and more admirable in its continuity and constancy than the living forest.

CAPTIONS

Designed by Heinrich Gohl, Basle

Plates by Interrepro AG, Reinach

Illustrations printed four-colour offset litho and
duplex offset litho by Graphische Betriebe Coop, Basle

Filmset by Siviter Smith & Co Ltd, Birmingham, England

Text printed offset litho by Fletcher & Son Ltd, Norwich, England

Text paper supplied by Frank Grunfeld Ltd, London

Bound by Richard Clay (The Chaucer Press) Ltd, Bungay, England